Rainfall - Observations and Modelling

Edited by Lakshmi Kumar TV and Humberto Alves Barbosa

Published in London, United Kingdom

Rainfall - Observations and Modelling
http://dx.doi.org/10.5772/intechopen.110919
Edited by Lakshmi Kumar TV and Humberto Alves Barbosa

Contributors
Bernard Tarza Tyubee, Daniel O. Omokpariola, Didier Gastmans, Hema Varssini Segar, John K. Nduka, Katleho Moloi, Michael Terver Iwan, Obiseye Obiyemi, Patrick L. Omokpariola, Puteri Natasha Sofia Zulkafli, Shuhaida Ismail, Vinicius dos Santos, Vladimir Eliodoro Costa, Zayra Christine Sátyro dos Santos

First published in London, United Kingdom, 2024 by IntechOpen
IntechOpen is the global imprint of INTECHOPEN LIMITED, registered in England and Wales, registration number: 11086078, 167-169 Great Portland Street, London, W1W 5PF, United Kingdom

British Library Cataloguing-in-Publication Data
A catalogue record for this book is available from the British Library

Additional hard and PDF copies can be obtained from orders@intechopen.com

Rainfall - Observations and Modelling
Edited by Lakshmi Kumar TV and Humberto Alves Barbosa
p. cm.
Print ISBN 978-1-83769-377-1
Online ISBN 978-1-83769-376-4
eBook (PDF) ISBN 978-1-83769-378-8

For EU product safety concerns:
IN TECH d.o.o., Prolaz Marije Krucifikse Kozulić 3, 51000 Rijeka, Croatia,
info@intechopen.com or visit our website at intechopen.com.

Meet the editors

Lakshmi Kumar TV is an associate professor in the School of Environmental Sciences, Jawaharlal Nehru University, New Delhi, India. Dr. Kumar's expertise is in climate change and land–atmospheric interactions. These areas include climate extremes and large-scale connections, the hydrological cycle and climate forcing agents, aerosol radiative forcing and its impact on the terrestrial hydrological cycle, etc. Dr. Kumar has published around 45 research papers in reputed journals and six book chapters.

Dr. Barbosa graduated in atmospheric sciences and received a master's degree in remote sensing from the National Institute for Space Research (INPE) and a Ph.D. in soil, water, and environmental sciences from the University of Arizona. Barbosa became a professor at the Federal University of Alagoas in 2006. He is the coordinator of the Satellite Image Analysis and Processing Laboratory (LAPIS—http://www.lapismet.com.br). He is a researcher in climate and ecosystem services. He uses remote sensing as an environmental monitoring tool, providing governments and business segments with satellite information. He participated as an author of a report for the Intergovernmental Panel on Climate Change (IPCC). Since 2008, Barbosa also collaborated with the Convection Working Group in the Meteosat satellites in detecting severe convective storms and nowcasting.

Contents

Preface

Rainfall, one of the key parameters of the earth's hydrological cycle, plays a vital role in the world's economy by providing livelihood for many people. Prediction of rainfall at various time scales attains significance in many places and will help the farmers and policymakers for agricultural use and for framing related policies, respectively. The use of traditional rain gauges for rainfall measurements paved the way for understanding regional variations of rainfall. The advent of the satellite era and the use of radar for rainfall strengthened numerical weather prediction, particularly the rainfall forecast. The climate models are very helpful for long-term rainfall prediction by offering a macroscale picture of the rainfall not only for historical periods but also for the future. The combined use of observations and modeling products provides a reliable picture of rainfall prediction.

There are many challenges in understanding the rainfall variation over centuries. Local conditions such as atmospheric instability and orographic lifting and global connections such as the El Nino southern oscillation (ENSO), Pacific decadal oscillation (PDO), North Atlantic oscillation (NAO), and so forth influence the rainfall seasonally and annually. The intra-seasonal and interannual variations of rainfall have not been understood completely due to the changing atmospheric dynamics across various regions. Human-induced climate change has affected the frequency of disasters and created devastating impacts. The increased frequency of droughts, floods, and heavy rainfall events challenges the scientific community to provide better predictions.

Developments in computational power and the use of artificial intelligence (AI) and machine learning (ML) complement the existing research on rainfall to simulate extreme weather events. The role of AI/ML is not only in the time series modeling of rainfall but also in grouping the climate models for better ensemble preparation among the given models. High computational power enables the rainfall forecast with finer time scales and corroborates the rainfall forecast during cyclonic times.

Keeping all the foregoing in view, the book Rainfall—Observations and Modeling invites scientific experts to submit their findings in the form of chapters. A thorough review took place in selecting the book chapters, and revisions were suggested before accepting papers for publication in the book. The chapters cover not only the previously mentioned areas but also areas of extended applicability of rainfall data. The editors of this book are very happy in working closely with IntechOpen, and the editors congratulate the authors of the book chapters for their excellent efforts in bringing out this book, which will be useful for post-graduate and doctoral students and early career researchers.

Lakshmi Kumar TV
School of Environmental Sciences,
Jawaharlal Nehru University,
New Delhi, India

Humberto Alves Barbosa
Institute of Atmospheric Sciences,
Federal University of Alagoas,
Alagoas, Brazil

Section 1

Applications of Rainfall Data in Meteorology

Chapter 1

Long Short-Term Memory Network versus Support Vector Machine for Flood Prediction

Hema Varssini Segar, Puteri Natasha Sofia Zulkafli and Shuhaida Ismail

Abstract

Malaysia is prone to flood disasters, which are considered the most hazardous natural disasters. This study compares the use of Long Short Term Memory (LSTM) networks and Support Vector Machines (SVM) in predicting future flash floods. Additionally, this study examines the effect of using the Synthetic Minority Oversampling Technique (SMOTE) in order to address imbalanced data. In this study, flooding for the year 2021 will be predicted based on the best-performing model. Experimental results indicated that the treatment had a positive impact on the study's outcome. An analysis of the outcomes of the models before and after treatment was conducted in order to determine which model delivers a higher degree of accuracy. SVM with RBF kernel is the most effective model before and after SMOTE treatment, out of all those evaluated in the study. Next, SVM model using RBF kernel after treatment was used to forecast flooding for 2021. Seven out of 12 floods were predicted by the model, which equates to 58.33% accuracy. Since the deep learning model did not perform well, future researchers could experiment with different numbers of hidden layers and hyperparameter settings to increase the accuracy.

Keywords: flood prediction, long short-term memory network, support vector machine, deep learning, synthetic minority oversampling techniques

1. Introduction

Natural disasters, such as flooding, have become the most frequent natural hazard, posing a threat to people as well as their possessions. People living in the affected regions had to cope with this situation. Consequently, they were forced to relocate to safer locations offered by the government, such as public buildings, institutions, and religious buildings [1]. According to the World Meteorological Organization [2], flooding can be classified according to its preceding events. The types of floods include flash floods, fluvial floods, seasonal floods, urban floods, and snowmelt floods.

Known as torrential floods [3], flash floods are among the most prevalent natural disasters, causing extensive damage both in urban and rural areas [4]. Torrential floods are typically characterized by roaring torrents that ravage riverbanks, roads, and mountain valleys, sweeping away anything in their path. Heavy rains can cause torrential floods within minutes or hours. Due to its speed, torrential floods are difficult to predict. This gives people a limited amount of time to flee or to bring food and other essentials with them.

Flood disasters occur quite frequently in Malaysia and have been named the most dangerous natural disaster alongside landslides, tsunamis, hurricanes, and haze [5]. Malaysia's Department of Irrigation and Drainage (DID) stated that flood prediction is intended to prepare an accurate and reliable forecast of imminent floods for citizens' safety while simultaneously reducing risks and property damage [6, 7]. The frequency of heavy rainfall leading to flooding in the past is expected to increase due to weather pattern changes. As a result, it is vital to be able to predict future flood events in order to minimize the damage caused by floods.

There are several techniques that can be used to develop flood prediction models, including machine learning and deep learning [7–9]. A number of studies have been conducted on flood disaster management and flood prediction systems. The goal of machine learning, a subfield of computer science, is to make it possible for computers to "learn" without being explicitly taught. Machine learning "learns" through becoming more skilled at predictions, classifications, and clustering through "practice" [9, 10].

In the context of classification and regression prediction, SVMs entail using machine learning techniques in order to enhance predicted accuracy without overfitting the data. There is a wide range of applications for SVM, including hydrology and meteorology. The reasons for this are due to the fact that hydrological and meteorological data are nonlinear [11], concluded that SVM was more effective at forecasting rainfall than statistical or numerical approaches. A study conducted by Hasan et al. [12] predicted rainfall in Bangladesh with 99.92% accuracy using a Support Vector Regression (SVR) model and concluded that SVR is better than traditional approaches. However [13], claimed that SVM has certain limitations, and one of them is the fact that its accuracy is highly dependent on parameters and kernel functions.

LSTMs refer to Long-Short-Term Memory networks that are modified forms of Recurrent Neural Networks (RNNs). LSTMs are often used to learn order dependence in sequence prediction problems, particularly in time series data. As a result of its superiority in the field of prediction, LSTMs are widely used in the field of hydrology [10]. LSTM was tested for low-flow forecasting in the Mahanadi River basin in India [14], and the most accurate outcome was found to be achieved by LSTM. The study concluded that LSTM is the best modern method and acts as an accurate Artificial Intelligence (AI) technique, especially for low-flow forecasting [15], conducted a study that utilized LSTM models to forecast discharge at the Hoa Binh Station on the Da River. Using LSTM models in hydrology may offer the possibility of developing and managing real-time flood warning systems.

A flood danger rating prediction model was developed by Kim and Kim [16] utilizing LSTM modeling and random forest methods. LSTM neural networks were used by Fang et al. [17] to predict flood susceptibility by combining an appropriate engineering method with an LSTM model. For preventing and mitigating flood hazards, researchers may be inspired by the Local Spatial Sequential Long-Short-Term Memory (LSS-LSTM) approach. Additionally, LSTM models have been successfully used in a variety of flood-related applications, such as precipitation forecasts, rainfall-runoff models, and river flood forecasts [10]. Although LSTM has difficulty displaying specific peak values, it follows dominant patterns, which is critical for flood prediction. It has been reported that [9] has discovered that deep-learning techniques such as Recurrent Neural Networks (RNNs) and LSTMs are being utilized to develop flood predictions using deep-learning methods. As Artificial Intelligence (AI) technology develops, an LSTM network model might be well suited to address hydrological engineering challenges due to its ability to model large time-series data using deep learning techniques. It has been confirmed that deep-learning methods provide high levels of accuracy [14–17].

In this study, we aim to develop supervised learning for flood prediction models based on the SVM algorithm and the LSTM network. The purpose of this study is to compare the accuracy of both models in predicting flooding in Subang Jaya. The SVM algorithm is a simpler prediction method, while LSTM networks are far more complicated. Understanding the pros and cons of each method will enable flood predictions to be made more accurately and efficiently in the future. This study could help future researchers choose between machine learning and deep learning when predicting upcoming floods.

2. Materials and methods

Explanation regarding data used in this project and the methods used to accomplish the objectives of this project are discussed in this section.

2.1 Data collection

Meteorological data alone especially using daily rainfall amount as the explanatory variable is seldom used to predict flooding, therefore there is not much info on flooding prediction using said data. The threshold of a variable is a crucial aspect when it comes to accuracies and precisely predicting flood. The city is in Peninsular Malaysia near Kuala Lumpur which is also susceptible to natural disasters such as flooding [18]. In this study, the meteorological dataset for the Subang Jaya area located at Sultan Abdul Aziz Shah Airport was obtained from the Malaysian Meteorological Department (MET Malaysia) located at Sultan Abdul Aziz Shah Airport. The data used in this project was 15 years' worth of meteorological data for Subang Jaya starting from 1st January 2005 until 31st December 2020. There are 5844 observations and 8 variables in this dataset. These variables are year, month, day, daily maximum temperature (°C), daily minimum temperature (°C), daily relative humidity (%), daily rainfall amount (mm), and daily mean sea level (MSL) pressure (hPa). The unavailability of data regarding the previous flooding incidents in Subang station from 1st January 2005 to 31st December 2020 has become a challenge to check the accuracy of the prediction model. Hence, there is room for error when constructing the flood prediction model. Apart from the lack of meteorological data, deep learning, and machine learning have not been compared side-by-side for flood prediction models.

2.2 Feature engineering

Data cleansing, data normalization, data transformation, and feature creation are all part of feature engineering [19]. In this research, data cleansing was done where missing data, outliers, and duplicate values were checked. Following feature creation, two distinct variables are developed: "rain intensity", as classified by Department of Meteorology, Malaysia in **Table 1**, and "flood occurrence", as classified by MET Malaysia in **Table 2**. Afterward, the training data was normalized.

Category	Minimum rainfall (mm)	Maximum rainfall (mm)
Slight rain	0	10
Moderate rain	10	60
Heavy rain	60	150
Extreme rain	150	∞

Table 1.
Categories of rainfall intensity according to its range.

Rain intensity	Flood occurrence
Slight rain	No
Moderate rain	No
Heavy rain	Yes
Extreme rain	Yes

Table 2.
Classification of flood occurrence based on rain intensity.

2.3 Data splitting

The data were split into two parts: training and testing data. The number of observations used as a training set is 4383 datapoints, starting from 1st January 2005 to 31st December 2016. For the remaining 1460 data points from 1st January 2017 to 31st December 2020 were utilized as a testing set. The training set was used to develop the model, while the testing set was used to measure the model accuracy and then to forecast future rainfall values.

2.4 Data balancing

Imbalanced data refers to a situation where data samples in a problem are not distributed equally. As a result, one or more classes in the dataset are undervalued. The uneven distribution of data causes the algorithm unable to perform forecasts accurately in predicting minority groups, resulting in various classification mistakes. The Synthetic Minority Oversampling Technique (SMOTE) was implemented to remedy the unequal distribution of unbalanced data [20]. SMOTE is a procedure where interpolation among neighboring minority classes is done. Without simply just replicating the minority class, SMOTE creates synthetic data by finding k-closer neighboring data [10].

2.5 Support vector machine (SVM) algorithm

Machine learning algorithms' expertise in data-driven learning rather than explicit teaching is also their main distinction from other computer programs. Numerous ML methods, such as Support Vector Machine (SVM), Naive Bayesian, and K-Nearest Neighbors (KNN) have proven to be crucial to support solutions in applications. SVM algorithm is a method of constructing a classifier that creates a decision boundary known as the hyperplane within the classes. Meanwhile, parameter tuning influences

Kernel	γ	C	d	coef0
Linear	N/A	1	N/A	N/A
RBF	0.2	1	N/A	N/A
Polynomial	0.2	1	3	0
Sigmoid	0.2	1	N/A	0

**N/A stands for not applicable.*

Table 3.
Hyperparameters setting for SVM models.

the classifier's efficiency and effectiveness [21]. Among others, the kernel function selected can have a significant impact on SVM model performance. However, trial and error are the best methods for selecting the best kernel. The models can be constructed beginning with a simple SVM and then experimented with various "standard" kernel functions [22]. **Table 3** shows the overall hyperparameters used in SVM models.

Apart from kernels, other hyperparameters used in the models are d = 3, γ = '*0.2*', *coef0* = 0, and C = 1 where all the values are default. The parameter 'degree' denoted as '*d*' is often used when dealing with a polynomial kernel reflecting the flexibility of a decision boundary. The parameter 'gamma' denoted as γ is the radius area for misclassification and the parameter "coefficient" is denoted as '*coef0*' significant only for polynomial and sigmoid kernel indicating the adjustable independent term in the kernel function. Finally, the parameter '*C*' adds a certain penalty for any misclassification of an observation. Each kernel function has its equations represented as the following:

$$Linear :< x,x' > \tag{1}$$

$$Polynomial : \left(\gamma < x,x' > +r\right)^{d} \tag{2}$$

$$RBF : \exp\left(-\gamma \|x,x'\| + r\right)^{d} \tag{3}$$

$$Sigmoid : \tanh\left(\gamma < x,x' > +r\right) \tag{4}$$

2.6 Long short-term memory (LSTM) network

The primary feature of an LSTM network is the hidden layer named as memory cells which has three gates named *forget gate* (f_t), *input gate* (i_t), and *output gate* (O_t). f_t gates are responsible for making decisions regarding what information is to be removed from the cell state. i_t gate is to specify what information ads to the cell state. O_t is to specify what information that was used from the cell state. The LSTM network has three major layers which are an input layer, single or multiple memory cells, and an output layer. The number of neurons in the input layer is equivalent to the number of the variables, in this would be 5. More layers require more computational time so, this project uses only two layers.

Hyperparameter	Value
Number of hidden layer	2
Activation	Hyperbolic tangent (tanh)
Recurrent_activation	sigmoid
Epoch	100 or 200
Batch size	64
Dropout	0.2
Optimizer	Stochastic Gradient Descent method (Adam)
Loss	binary_crossentropy

Table 4.
Hyperparameter setting for LSTM (2) model.

A memory cell starts functioning by removing information from the previous cell state, S_{t-1} are determined by the LSTM layer. At time stamp *t*, the activation value f_t of the forget gates is calculated using the existing input, x_t and output, h_{t-1} of the memory cells at $t-1$ and the bias terms of forgetting gates, b_f. The sigmoid function adjusts all activation levels to 0 and 1 as entirely forgotten and completely recalled, respectively. Secondly, the LSTM layer would determine the information that was included in the network's cell states, S_t. This computes the candidate's values, $\widetilde{S}_t$ and then activation values of the input gates. Thirdly, where new cell states, S_t are calculated using the information of the preceding 2 steps. Finally, the output, h_t of the memory cells unit is acquired. The hyperparameters used in this typical LSTM model include certain functions with default values and are listed in **Table 4** [23]. The activation function enables the model to undergo non-linear processes.

Recurrent_activation is used for the recurrent step and optimizer is used to reduce the overall loss and improve accuracy. The optimizer used for this model is the stochastic gradient descent method known as *'Adam'* which is based on adaptive estimation of first-order and second-order moments. Loss identifies if the prediction is good (low loss value) or bad (high loss value).

2.7 Confusion matrix, receiver operating characteristics (ROC) curve and area under the curve (AUC) value

To evaluate the performance of the models, the confusion matrix was constructed since it provided information such as accuracy, precision, recall, F1-Score, and the AUC for the ROC curve [24]. **Table 5** shows the framework for a confusion matrix.

		Actual class	
		Positive	**Negative**
Predicted class	Positive	True Positive	False Positive
	Negative	False negative	True negative

Table 5.
Confusion matrix.

The equations used to calculate the Precision, Recall, F1-Score, Accuracy, and Area Under the Curve (AUC) values are as follows:

$$Accuracy: A = \frac{TP + TN}{TP + FP + TN + FN} \tag{5}$$

$$Precision: P = \frac{TP}{TP + FP} \tag{6}$$

$$Recall: R = \frac{TP}{TP + FN} \tag{7}$$

$$F\_1 - score: F = \frac{(2 * P * R)}{(P + R)} \tag{8}$$

$$AUC = \frac{\Sigma TP + \Sigma TN}{(P + N)} \tag{9}$$

3. Results and discussion

Data analysis of the output from the experiments is shown in this section. The results from the models constructed discussion and analyzed by using the confusion matrix, ROC curve, and AUC. This section also discusses the objectives of this research.

3.1 Feature engineering

There are a number of steps discussed under this subtopic, including handling missing data, checking for outliers, creating features, and normalizing data. The column daily relative humidity has 10 missing values. As the number of missing values was less than 5%, they are not considered significant. Therefore, imputation was performed by replacing missing values with the mean of the column daily relative humidity (%), which is 78.13. Apart from missing values, no duplicate dates were found for this dataset. **Figure 1** shows a boxplot constructed to check for outliers. The maximum temperature, humidity, and rainfall are the only three variables that exhibit outliers. Meanwhile, the minimum temperature has a slight negative skewness.

Next, variables minimum temperature and rainfall are skewed positively. Both, variables humidity and MSL pressure on the other hand show normal distributions. Since coming across an extreme value is common for meteorological data, it is determined that outliers are normal to exist and should not be removed [25]. After cleaning the data, feature creation was done where two new variables were created using daily rainfall amount. The variables created were based on information in **Tables 1** and **2** as viewed in Section 2. Normalizing data is an important part of data pre-processing to ensure that the scale has been changed without distorting the information of an observation. In this research, the dataset was normalized in the range 0 to 1. This maintains the general distribution of the observation but simply converts the data to improve performance and accuracy.

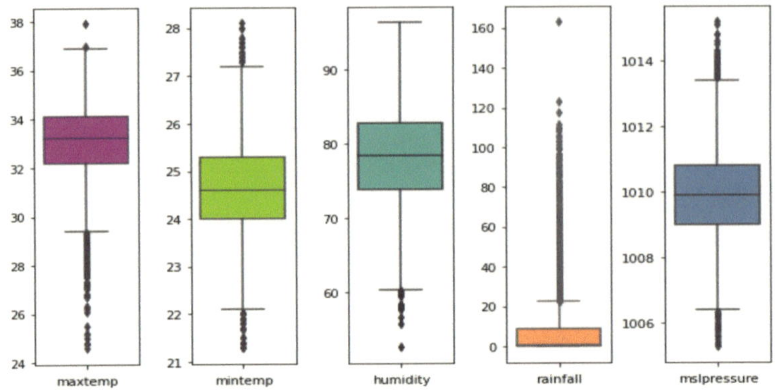

Figure 1.
Boxplot for each variable.

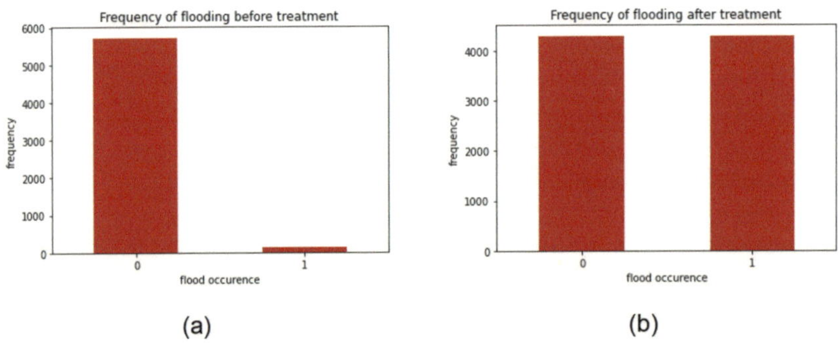

Figure 2.
(a-b) Bar chart of balance of class for flood occurrences before and after treatment.

3.2 Data balancing

Without treatment, the accuracy and precision value might seem, high, however, the prediction is fundamentally biased in the sense of the representation of classes. **Figure 2(a)** shows the frequency of classes before treatment. The bar chart shows that the classes are severely imbalanced and would require treatment to ensure unbiased predictions. "1" is denoted as "flooding" and "0" is denoted as "no flooding," **Figure 2(b)** shows the bar chart after the treatment. The classes are balanced where more "1" is predicted by producing synthetic data using SMOTE. Prediction models using both before and after treatment are done for both machine and deep learning methods and the difference of accuracy is shown.

3.3 Support vector machine (SVM)

The SVM algorithm was used to construct a total of 4 models before and after treatment, respectively. This is because each model is constructed using different hyperparameter settings. By constructing the models using the suitable hyperparameter settings, the outputs were obtained. **Table 6** shows the results from the confusion matrix for all SVM models using different hyperparameter settings before and after treatment. According to **Table 6**, SVM models using linear kernels has performed

Kernel	Treatment	Precision	Recall	F1-score	Accuracy	AUC
Linear	After	1.0000	1.0000	1.0000	1.0000	1.0000
	Before	1.0000	1.0000	1.0000	1.0000	1.0000
RBF	**After**	**0.9758**	**0.8790**	**0.9217**	**0.9945**	**0.8790**
	Before	0.9480	0.6721	0.7465	0.9863	0.6721
Polynomial	After	0.6908	0. 9836	0. 7678	0.9678	0.9836
	Before	0.9911	0.5517	0.5892	0.9822	0.5517
Sigmoid	After	0.0099	0.5000	0.0195	0.0199	0.5000
	Before	0.4901	0.5000	0.4950	0.9801	0.5000

**Bold values indicate best-performing model.*

Table 6.
SVM results for different hyperparameter settings before and after treatment.

Evaluation metrics	Before treatment	After treatment	Percentage of improvement (%)
Precision	0.9480	0.9758	2.85
Recall	0.6721	0.8790	23.54
F1-Score	0.7465	0.9217	19.01
Accuracy	0.9863	0.9945	0.82
AUC	0.6721	0. 8790	23.54

Table 7.
Percentage of improvement SVM models using RBF kernels before and after treatment.

very well, however, the perfect results using linear kernel exhibits signs of overfitting. Overfitting models are highly prone to provide misleading results.

Furthermore, the experiments showed that SVM with RBF have the most feasible results for both before and after SMOTE treatment. Before treatment, the AUC value is 0.6721 and the accuracy is 0.9863 meanwhile after the treatment, the AUC value increases to 0.8790 and the accuracy to 0.9945. SVM with RBF also showed stability in terms of statistical performance measurements, compared to other kernel functions.

Table 7 shows the percentage of improvement for SVM with RBF kernel after SMOTE treatment. Detailed analysis showed the precision improved by 2.85% from 0.9480 to 0.9758 and accuracy has improved from 0.6721 to 0.8790 which is 23.54%. Overall, all statistical performance measurement exhibit an increased performance after SMOTE treatment.

When it comes to ROC curve, a steeper curve indicates a better-performing model. **Figure 3(a-b)** shows ROC curves of SVM models each using RBF kernel and its specified hyperparameter settings before and after treatment which exhibits that the best performing kernel is the RBF kernel since there is no overfitting yet feasible result. **Figure 3(a)** reveals a moderately performing ROC curve with AUC value of 0.6721. However, there is still room for improvement in terms of the accuracy of the models. **Figure 3(b)** shows the ROC curves that performed better, and the ROC curve has a value of 0.8790 which supports the claim that treatment for SVM algorithm of RBF kernel is the best model.

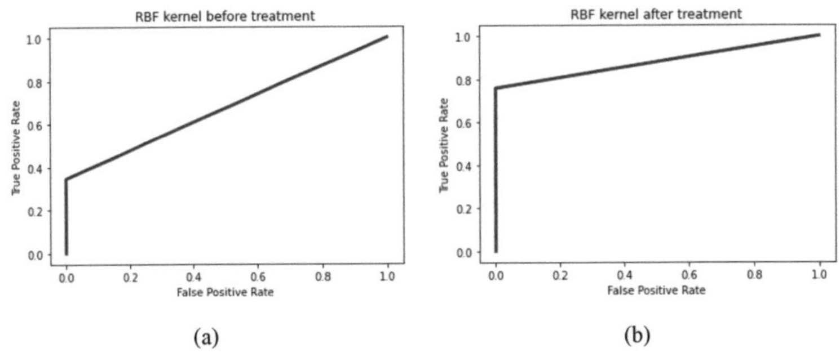

Figure 3.
(a-b) ROC Curve for SVM algorithm RBF kernel before and after treatment.

3.4 Long short-term memory (LSTM) network

This research utilizes an LSTM network with a total of 4 layers constituting 1 input layer, 2 hidden layers, and 1 output layer. The input and hidden layers have 5 nodes each and the output layer has 1 node. This can be changed accordingly through trial and error. A summary of the output obtained from the confusion matrices is in **Table 8**. The best-performing model according to **Table 8** is LSTM with 100 epochs before treatment due to the highest AUC value that is feasible.

Table 9 was constructed to better view the percentage of improvement for best best-performing LSTM model before and after treatment. **Table 9** proves to apply treatment of SMOTE to the LSTM of epoch 100 prediction model has decreased its prediction ability. The precision of model has decreased by 23.85% from 0.7640 to 0.6169 and the accuracy has decreased from 0.9822 to 0.9349 which is a 5.06% deterioration. Finally, the AUC has increased slightly (1.00%) from 0.9571 to 0.9668. Overall, treatment has not shown much difference for the model.

Epoch	Treatment	Precision	Recall	F1-score	Accuracy	AUC
Epoch = 100	After	0.6169	0. 9668	0. 6724	0.9349	0.9668
	Before	0.7640	0. 9571	0. 8329	0.9822	0.9571
Epoch = 200	After	0.6466	0.9266	0.7106	0.9555	0.9752
	Before	0.7290	0.9040	0.7905	0.9774	0.9195

Table 8.
LSTM results for different hyperparameter settings before and after treatment.

Evaluation metrics	Before treatment	After treatment	Percentage of improvement (%)
Precision	0.7640	0.6169	−23.85
Recall	0.9571	0.9668	1.00
F1-Score	0.8329	0.6724	−23.87
Accuracy	0.9822	0.9349	−5.06
AUC	0.9571	0. 9668	1.00

Table 9.
Percentage of improvement LSTM model for 100 epochs before and after treatment.

Figure 4(a) and **(b)** show the plot of loss versus accuracy for the constructed LSTM (2) networks before and after treatment. **Figure 5(a)** and **(b)** shows the plot have accuracy and loss that are a good fit continuously likely leading to an overfit. **Figure 5(c)** and **(d)** shows the loss and accuracy converged to a certain epoch value and diverged to maximum value meaning the models had trained well for the dataset and produced relatively high accuracy.

3.5 Forecasting flood for year 2021

The best-performing model has been chosen as the SVM algorithm model utilizing the RBF kernel after treatment since it has a higher precision and accuracy which is 0.9758 and 0.9945 respectively compared to LSTM of 100 epochs after treatment which has a precision and accuracy of 0.6169 and 0.9349 respectively. Since the best model is determined as SVM model of RBF kernel after treatment, the forecasting of flooding was done using raw meteorological data for the year 2021. The expected output was Subang Jaya should have experienced flooding for a total of 12 days in the year 2021. However, the outcome was that Subang Jaya experienced 7 days of flooding only. The ratio of outcome to expected outcome is 7:12 which is approximately 58.33%. The model failed to predict flooding 41.67% of the times. This might be due to certain factors such as temperature, humidity, and MSL pressure fluctuations that were not detected by the model.

Figure 5 shows the ROC curves for LSTM models of respective hyperparameter settings before and after treatment is given. All ROC curves for the LSTM model of different epochs have performed well. However, the best-performing model is LSTM of epoch 100 after treatment since the ROC curve is steep and almost reaches the upper left corner of the graph. This claim is supported by the AUC value of 0.9668 proving that LSTM of 100 epoch after treatment performed the best.

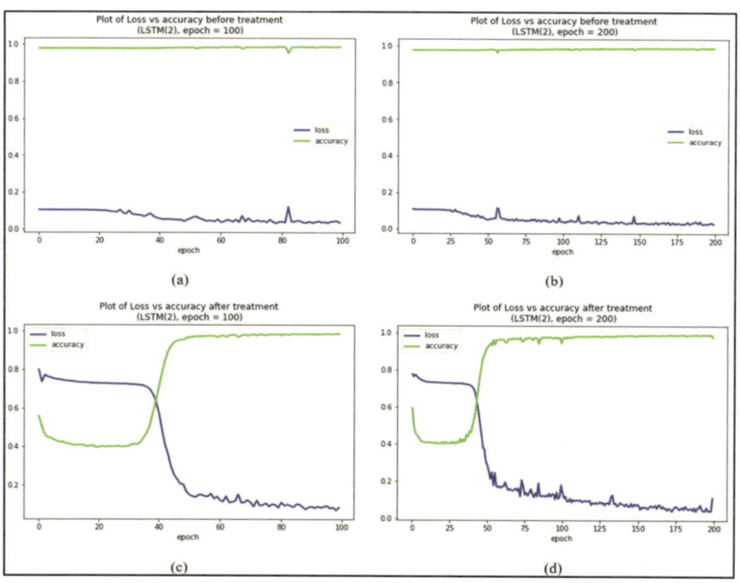

Figure 4.
(a-d) Plot loss vs. accuracy for LSTM of different epochs before and after treatment.

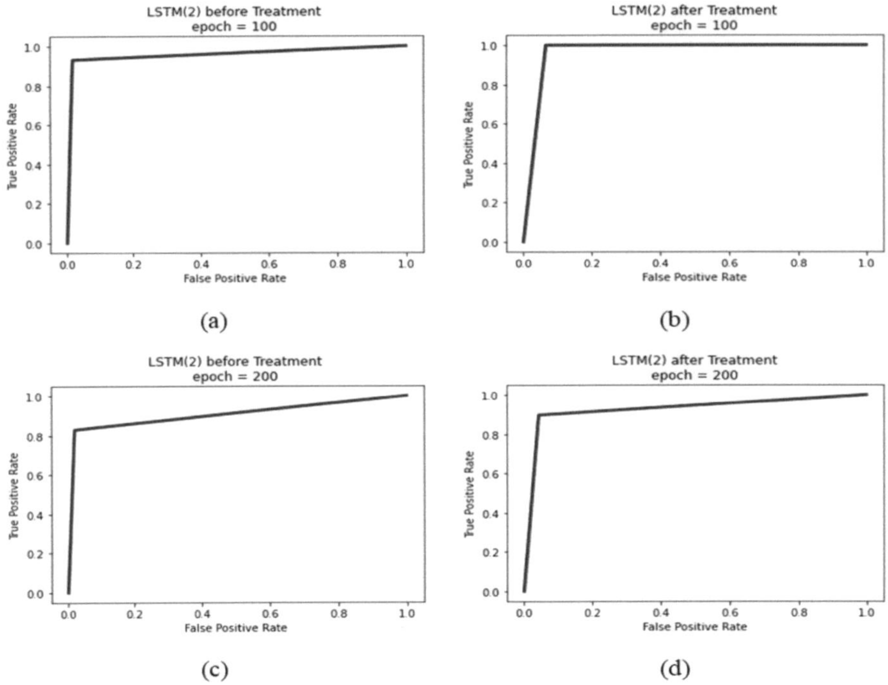

Figure 5.
(a-d) ROC for LSTM of different epochs before and after treatment.

4. Conclusion

To conclude, the SVM model with RBF kernel, γ = '0.2,' and C = 1 after treatment would be the appropriate choice to construct a future flood prediction model based on meteorological data for the Subang Jaya area. This is because SVM models require lesser computational time and are easier to compute compared to LSTM networks since SVM algorithms do not need multiple layers. Besides, the accuracy for the SVM model using RBF kernel after treatment is 0.9945 which is exceptionally good. Using SMOTE as a treatment for imbalanced data has also shown significantly better accuracy in the models. The findings from this study indicate that SVM is better compared to LSTM. Due to the fact that SVM models require less computational time and are simpler to compute than LSTM networks. SVM algorithms do not require multiple layers of computation. Additionally, the accuracy of the SVM model using RBF kernel after treatment is 0.9945, which is exceptionally high. The use of SMOTE as a treatment for imbalanced data has also demonstrated a significant improvement in the accuracy of the models.

Future researchers will be able to take into account the computational time and complexity of the model when predicting flooding using meteorological data. The positive impact of using SMOTE in this research will also assist researchers in developing a better flood prediction model in the future, preventing biased results from occurring. To improve the accuracy of the models, future researchers may utilize different values of hyperparameters. It would be possible to produce a better prediction using LSTM network by adding more hidden layers and nodes within the layers. More hidden layers and nodes might help the neural network to learn more features and predict accurately. Higher epochs and batch size might also help increase the accuracy of a neural network. As for the SVM models, more hyperparameters could be explored to produce better predictions.

Acknowledgements

This research was supported by research grant FRGS/1/2022/ICT06/UTHM/03/1 provided by the Ministry of Higher Education, Malaysia. The authors would also like to thank the Faculty of Applied Sciences and Technology, Universiti Tun Hussein Onn Malaysia for its support.

Conflict of interest

The authors declare no conflict of interest.

Data availability statement

Data is deemed private and confidential and will not be shared.

Author details

Hema Varssini Segar, Puteri Natasha Sofia Zulkafli and Shuhaida Ismail*
Universiti Tun Hussein Onn Malaysia, Pagoh Edu Hub, Johor, Malaysia

*Address all correspondence to: shuhaida@uthm.edu.my

References

[1] Saimi FM, Hamzah FM, Toriman ME, Jaafar O, Tajudin H. Trend and linearity analysis of meteorological parameters in peninsular Malaysia. Sustainability. 2020;**12**(22):9533-9552

[2] World Meteorological Organization. Manual on Flood Forecasting and Warning. Switzerland: Publications Board World Meteorological Organization (WMO); 2020

[3] Bi Q, Goodman KE, Kaminsky J, Lessler J. What is machine learning? A primer for the epidemiologist. American Journal of Epidemiology. 2019;**188**(12):2222-2239

[4] Syeed MMA, Farzana M, Namir I, Ishrar I, Nushra MH, Rahman T. Flood prediction using machine learning models. In: 2022 International Congress on Human-Computer Interaction, Optimization and Robotic Application (HORA). 2022. pp. 1-6

[5] Adib M, Razi M, Tahir W, Alias N, Ismail LH, Ariffin J. Development of rainfall model using meteorological data for hydrological use. International Journal of Integrated Engineering. 2013;**5**(1):64-73

[6] Department of Irrigation and Drainage. National Flood Forecasting and Warning Centre. 2022. Available from: https://www.water.gov.my/index.php/pages/view/195 [Accessed: May 15, 2022]

[7] Mosavi A, Ozturk P, Chau KW. Flood prediction using machine learning models: Literature review. Water (Switzerland). 2018;**10**(11):1536-1548

[8] Faruq A, Arsa HP, Hussein SFM, Razali CMC, Marto A, Abdullah SS. Deep learning-based forecast and warning of floods in Klang River, Malaysia. Ingenierie Des Systemes d'Information. 2020;**25**(3):365-370

[9] Moishin M, Deo RC, Prasad R, Raj N, Abdulla S. Designing deep-based learning flood forecast model with ConvLSTM hybrid algorithm. IEEE Access. 2021;**9**:50982-50993

[10] Kratzert F, Klotz D, Brenner C, Schulz K, Herrnegger M. Rainfall- runoff modelling using long short-term memory (LSTM) networks. Hydrology and Earth System Sciences. 2018;**22**(11):6005-6022

[11] Hirani D, Mishra N. A survey on rainfall prediction techniques. International Journal of Computer Application. 2016;**6**(2):28-42

[12] Hasan N, Nath NC, Rasel RI. A support vector regression model for forecasting rainfall. In: 2nd International Conference on Electrical Information and Communications Technologies (EICT). 2015. pp. 554-559

[13] Zhang J, Qiu X, Li X, Huang Z, Wu M, Dong Y. Support vector machine weather prediction technology based on the improved quantum optimization algorithm. Computational Intelligence Neuroscience. 2021:1-13. DOI: 10.1155/2021/6653659

[14] Sahoo BB, Jha R, Singh A, Kumar D. Long short-term memory (LSTM) recurrent neural network for low - Flow hydrological time series forecasting. Acta Geophysica. 2019;**67**(5):1471-1431. DOI: 10.1007/s11600-019-00330-1

[15] Le X, Ho H. V, Lee G, Jung S. Application of long short-term memory (LSTM) neural network for flood forecasting. Water. 2019;**11**(7):1387

[16] Kim HI, Kim BH. Flood Hazard rating prediction for urban areas using random Forest and LSTM. KSCE Journal of Civil Engineering. 2020;**24**(12):3884-3896. DOI: 10.1007/s12205-020-0951-z

[17] Fang Z, Wang Y, Peng L, Hong H. Predicting flood susceptibility using LSTM neural networks. Journal of Hydrology. 2021;**594**:125734. DOI: 10.1016/j.jhydrol.2020.125734

[18] Rahman IA, Dewsbury J. Selection of typical weather data (test reference years) for Subang, Malaysia. Building and Environment. 2007;**42**(10):3636-3641

[19] Bekar ET, Nyqvist P, Skoogh A. An intelligent approach for data pre-processing and analysis in predictive maintenance with an industrial case study. Advances in Mechanical Engineering. 2020;**12**(5):1-14

[20] Asniar, Maulidevi NU, Surendro K. SMOTE-LOF for noise identification in imbalanced data classification. Journal of King Saud University - Computer and Information Sciences. 2021;**34**(6):3413-3423

[21] Ramasamy LK, Kadry S, Nam Y, Meqdad MN. Performance analysis of sentiments in twitter dataset using SVM models. International Journal of Electrical and Computer Engineering. 2021;**11**(3):2275-2284

[22] Huang S, Nianguang CAI, Penzuti Pacheco P, Narandes S, Wang Y, Wayne XU. Applications of support vector machine (SVM) learning in cancer genomics. In Cancer Genomics and Proteomics. 2018;**15**(1):41-51

[23] Team, K. Keras Documentation: LSTM layer. Keras.io. 2023. Available from: https://keras.io/api/layers/recurrent_layers/lstm/. [Accessed: January 11, 2023]

[24] Sellami EM, Maanan M, Rhinane H. Performance of machine learning algorithms for mapping and forecasting of flash flood susceptibility in Tetouan, Morocco. The International Archives of the Photogrammetry, Remote Sensing and Spatial Information Sciences. 2021;**2022**(4):305-313

[25] Zhang AZ, Li JZ, Gao H, Chen YB, Ma HZ, Bah MJ. CrowdOLA: Online aggregation on duplicate data powered by crowdsourcing. Journal of Computer Science and Technology. 2018;**33**(2):366-379

Chapter 2

Spatio-Temporal Analysis of Dry and Wet Spells in the Middle Belt of Nigeria

Bernard Tarza Tyubee and Michael Terver Iwan

Abstract

The spatial patterns and trends of various categories of dry and wet spells were analysed from 1981 to 2010 in the Middle Belt of Nigeria. Daily rainfall (mm) data were obtained from eight synoptic weather stations spread across the region. The spatial variation and temporal trend of spells were analysed using the coefficient of variation (CV) and Pearson's correlation coefficient (r). The result reveals that spatially, dry spells varied from 12.8 to 110.1%, while wet spells varied from 11.7 to 192.5%. The longest dry spell length by station ranged from 14 days (Jos) to 37 days (Yola), while the longest wet spell ranged from 7 days (Bida, Ibi and Makurdi) to 11 days (Ilorin and Jos). Both dry and wet spells exhibited positive and negative trends. Significant trends of dry spells include negative trends of categories 2–4 days (Ibi), 8–10 days (Yola), 11 days+ (Ilorin and Yola); and positive trends of categories 2–4 days (Lokoja) and 8–10 days (Ilorin). For wet spells, only positive trends were significant. The study concludes that the south western (northern) part of the region recorded the highest (least) annual frequency of dry spells and least (highest) annual frequency of wet spells, respectively.

Keywords: wet spells, dry spells, trend, variation, patterns

1. Introduction

The study of daily rainfall occurrence, especially dry and wet spells, is of high climatologic interest across the world. During the past few decades, rainfall distribution in sub-Saharan Africa has been observed to have high spatial and temporal variability [1]. Dry and wet spells are two main physical characteristics of rainfall, the distribution of which influences the volume of rainfall in a geographical area [2–5]. Also, events of dry and wet spells have a critical influence on the duration of rainy season, mostly, in terms of false onset or early or late cessation of the season [6]. Thus, rain spells have been a topical issue of increasing importance globally. Global climate models have indicated a rising frequency of summer droughts due to global warming, even in regions where droughts were previously rare events, while at the same time, devastating floods in Nigeria, Europe and other parts of the world have increased the interest in wet spells and heavy precipitation occurrences [7–11].

Moreover, the persistence properties of the daily precipitation process are governed by spell lengths. Therefore, spells of dry days can reveal significant changes in the structure of drought, while identifying changes in the trend of both dry and wet spells as well as their persistency will provide useful information in predicting future climate events since these variables are closely related to extreme weather events such as drought and flood [12, 13].

The distribution of dry and wet spells remains an important applicative approach to rainfall statistics as it provides useful knowledge and information for many areas particularly agriculture [14–18], irrigation and water resource management [19, 20]. Knowledge of the period of occurrences, frequencies and duration of dry and wet spells will minimise unexpected damage due to drought or flood. Information on rainfall probabilities is also vital for the design of water supply management, supplementary irrigation schemes and the evaluation of alternative cropping systems for effective soil water management plans [21]. Such information can also be beneficial in determining the best-adapted plant species and the optimum time for seedlings to re-establish vegetation on deteriorated rangelands. Consequently, several studies were focused on dry (wet) spells since these variables greatly influence agriculture, drought and flood [10, 13, 22–25].

Synoptically, rain spells are associated with large-scale atmospheric circulation patterns and coincide almost always with the presence of specific meteorological features such as depressions, cyclonic systems, fronts and troughs, whereas the absence of these phenomena is almost tantamount to the absence of rain [26–28]. For instance, the 2–3-, 4–5- and 6–9-day wet spells are reported to be associated with the propagation of the African Easterly Waves, while the 10–20-day wet spells are linked to the coupled land-atmosphere interaction in the African Monsoon rainfall [1, 29]. On the other hand, lack of rainfall (dry spell) is associated with anticyclone systems that can last from 1 day to many days or semi-permanent or seasonal anticyclones [30]. Therefore, knowledge of the characteristics of rain spells, from the synoptic view point, is relevant in recognising rain-bearing characteristics of these weather systems, while the rain spells themselves appear a more "natural" time unit for analysis of rainfall than the conventional ones [26].

Like other tropical regions, rainfall in the Middle Belt region and Nigeria at large, greatly varies temporally and spatially. Its onset and cessation, frequency, intensity, amount and sequences of spells also vary greatly. While rainfall characteristics such as onset, cessation, duration, intensity, rain days, seasonality, amount, persistence and periodicity are well researched and documented in the Middle Belt region [31–34], studies on spells are rather scarce. A recent study in the region on dry and wet spells [15] focused on their frequency distribution and probabilities based on the Markov chain model, to guide agricultural cop planning in the region. Another more recent study [18] on dry spell prediction covered the whole of Nigeria, with nine study stations of which three (Ilorin, Lokoja and Makurdi) fall within the Middle Belt region. The focus of this study was on the prediction of critical dry spells in rain-fed maize crop production for various locations using an artificial neural network.

However, in addition to agricultural production, knowledge of dry and wet spells in the Middle Belt region is crucial to hydrologists, environmentalists, agriculturists, planners and climate-induced disaster managers. This chapter, therefore, analyses the spatial patterns and trends of both dry and wet spells for the Middle Belt region. The findings of this work are relevant in serving the information needs of the aforementioned interest groups and also adding to the existing knowledge and literature on rainfall climatology of the region.

2. Data collection, definition and categorisation spells

2.1 Data collection

Daily rainfall (mm) data were obtained from the Nigerian Meteorological Agency (NiMet) eight synoptic weather stations spread across the Middle Belt region for a 30-year period (1981–2010). The stations are Bida, Ibi, Ilorin, Jos, Lokoja, Makurdi, Minna and Yola. The choice of the stations was due to data quality and availability for the study period, and geographical spread.

2.2 Definition of spells

A spell is defined, in the study, as a consecutive number or group of days each with at least 1.2 mm of rainfall (wet spell) or less than 1.2 mm of rainfall (dry spell). As Hern'aeza and Martin-Vide [35] noted, the definition of thresholds is important because it plays a key role later on in the outcomes of the study. Fischer et al. [36] specifically pointed out that for agricultural applications a threshold of 1 mm day^{-1} may be more appropriate. This study, however, uses 1.2 mm as a threshold for description of dry/wet days, which translates to a weekly threshold of 8.4 mm. These thresholds relate more closely to the weekly crop water requirement for Nigeria, which is 8 mm [37, 38]. The 1.2 mm daily threshold also takes care of measurement errors associated with light rains due to direct evaporation at manual rain gauges [39, 40]. Thus, a day is reckoned as wet if it has rainfall equal to or more than 1.2 mm, while days with less than 1.2 mm are considered as dry. Similarly, a wet week is one with rainfall equal to or more than 8.4 mm, and a dry week has less than 8.4 mm.

2.3 Categorisation of spells

The dry and wet spells are categorised into four and two classes, respectively (**Table 1**). Though dry and wet spells vary in duration, the categories of dry spells are chosen with reference to the water stress on various crop types caused by dry spells of varying durations [41], and those of wet spells are chosen to correspond to the different synoptic systems causing rain in West Africa. The wet spells lasting 2–4 days are associated with the so-called "3–5 days" African Easterly Waves (AEWs), while those lasting for 5 days and above are related to the "6–9 days" African Easterly Waves [29].

Duration (days)	Spell	
	Dry	Wet
2–4	✓	✓
5+		✓
5–7	✓	
8–10	✓	
11+	✓	

Table 1.
Classification of dry and wet spells.

3. Determination of onset and cessation of the rainy season

Gitau [42] made a useful suggestion that the actual onset and cessation dates of the rainfall season should be determined before the frequency distribution of wet and dry spells is derived. Since the scope of this study is limited to the rainy season, onset and cessation of the season have been defined and empirically determined from daily rainfall data. Based on Tarhule and Woo [39] method, the onset of the rainy season is taken to be the first wet day of the year after which there is no dry spell longer than 12 days in-between the subsequent 4 wet days, while the dry season is considered terminated on any wet day after which there occurred a dry spell longer than 12 days during the last 4 wet days of the year. The condition of having no dry spell of more than 12 days in-between 4 consecutive wet days following the onset day is to eliminate the possibility of including a false onset of the rainy season. The duration of the season is the number of days between the onset and cessation dates.

To determine the onset date of the rainy season for a given year, the daily rainfall data were observed from the beginning of the year to identify a wet day, then the next 4 consecutive wet days were also identified, ensuring that no dry spell of duration longer than 12 days occurred between them. The first wet day after which no dry spell longer than 12 days occurred, before the next 4 consecutive wet days were counted, marked the onset of the rainy season. On the other hand, cessation of the season was determined by observing the daily rainfall data backward from the end of the year to identify the wet day after which a dry spell longer than 12 days occurred during the last 4 wet days of the year.

4. Results

4.1 Spatial variation in dry and wet spell

To analyse the spatial variation of spells in the Middle Belt (MB) of Nigeria, isolines and coefficient of variation (CV) of the annual frequency of spells from 1981 to 2010 were used. The analysis was based on categories of dry and wet spells.

4.1.1 Spatial variation in dry spells

The total annual frequency of 2–4-day dry spells decreases from the southern to the northern part of the region. However, Lokoja and Jos recorded the highest and lowest annual frequency of 682 and 510 spells (**Figure 1a**). Makurdi (with 649 spells) comes second after Lokoja in the order of high occurrences but lies within the same zone with Bida and Ilorin where the occurrence number is between 600 and 650 spells. There is an enclave of low occurrences towards the eastern and north eastern parts of the region where Ibi and Yola have 552 and 534 spells, respectively, while Minna in the north western part is isolated with 596 spells. The CV values showed that Jos has the least station variation (CV = 13.4%), while Ibi has the largest station variation with CV values of 26.7%, respectively (**Table 1**).

For the 5–7 days category, all the Niger-Benue valley areas of Lokoja, Makurdi, Ibi and Yola, plus Ilorin in the south west fall within the same zone with a total number of occurrences between 150 and 200 dry spells (**Figure 1b**). There is a decrease in the annual frequency of occurrence from this zone towards the north central plateau where, again, the lowest value is at Jos, with a total annual occurrence of 83 dry spells.

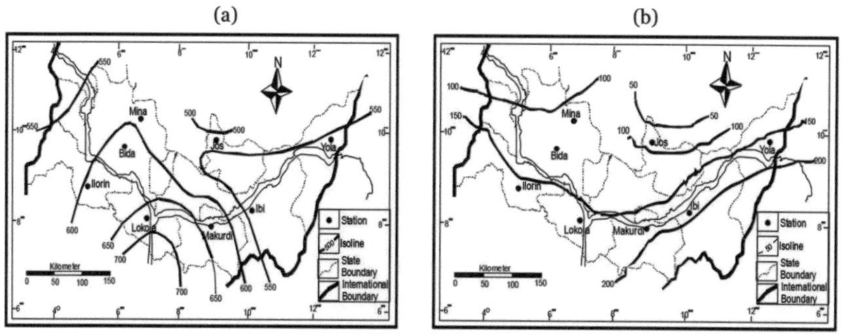

Figure 1.
Spatial variation annual frequency of 2–4-day (a) and 5–7-days (b) dry spells in MBR.

The CV values, however, showed that Ilorin has the least station variation (CV = 33.5%), while Jos has the largest station variation with CV values of 67.6%, respectively (**Table 1**).

The spatial pattern of annual frequency of 8–10-day dry spells category showed a zonal pattern with an increase from the southern part, along Ilorin-Lokoja-Ibi axis to the northern part (Jos). Similarly, Ilorin and Lokajo recorded the highest annual total of 50 and 60 dry spells, whereas Jos recorded the least frequency of 18 dry spells (**Figure 2a**). The CV values showed that Lokoja has the least station variation (CV = 55.8%), while Jos has the largest station variation with CV values of 120.7%, respectively (**Table 1**).

The spatial pattern of annual frequency of 11 days+ dry spells (**Figure 2b**) is similar to the 8–11-day dry spells (**Figure 2a**). However, the annual total frequency was lower relative to 8–10-day dry spells with the highest and least annual frequency of 38 and 11 dry spells observed in Ilorin and Jos/Minna, respectively (**Figure 2b**). For station-to-station variation, the CV values showed that Ibi has the least station variation, with a CV value of 72.3%, while Ibi has the largest station variation with CV values of 182.4%, respectively. Generally, it can be inferred from the spatial distribution pattern that the annual frequency of the four categories of dry spells was higher in the southern part and lower in the northern part when compared to the frequency of the average region (**Table 2**).

The result suggests that the annual number of dry spells decreases from the southern to the northern part of the region. However, the annual frequency and CV of dry

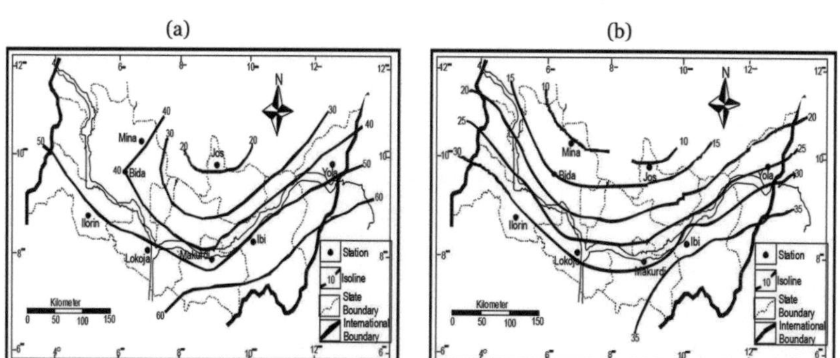

Figure 2.
Spatial variation in annual frequency of 8–10-day (a) and 11-day+ (b) dry spells in MBR.

Station	Duration of spell											
	2–4 days			5–7 days			8–10 days			11 days+		
	Total	Mean	CV	Total	Mean	CV	Total	Mean	CV	Total	Mean	CV
Bida	617	21	18.2%	138	5	50.0%	40	1	77.1%	15	1	114.5%
Ibi	534	18	26.7%	199	7	39.3%	53	2	65.9%	34	1	72.3%
Ilorin	601	20	21.8%	153	5	33.5%	56	2	81.8%	38	1	82.8%
Jos	510	17	13.9%	83	3	67.6%	18	1	120.7%	11	0	182.4%
Lokoja	682	23	21.5%	155	5	41.9%	56	2	55.8%	26	1	94.5%
Makurdi	649	22	15.4%	155	5	42.6%	42	1	85.1%	28	1	92.1%
Minna	596	20	16.4%	112	4	45.6%	44	1	114.3%	11	0	151.7%
Yola Region	552	18	17.1%	152	5	43.0%	43	1	74.8%	24	1	100.6%
	593	20	18,20%	143	5	44.18%	44	1	82.27	23	1	111.22%
CV Range			12.8%			34.1%			64.9%			110.1%

Table 2.
Descriptive statistics of dry spells in the Middle Belt of Nigeria (1981–2010).

spells have increased from shorter duration (2–4 days) to longer duration (11 days+) with Jos recording the least annual frequency and least station variation for all the categories of dry spells.

4.1.2 Spatial variation in wet spells

Spatial variation of wet spells in the region showed that there is an increase in the total annual frequency from south to north. For the 2–4-day category, Jos (north) and Lokoja (south) have the highest and least number of 2–4-day wet spells of 557 and 329 spells, respectively (**Figure 3a**), whereas Jos (north) and Ilorin (south) have the highest and least annual frequency of 100 and 40 spells for the 5-day+ category (**Figure 3b**). The results of coefficient of variations of the various spell categories (**Table 3**) showed that the CV ranges from 17.3% (Ilorin) to 27.9% (Ibi) for 2–4-day wet spells and from 31.7% (Jos) to 224.2% (Yola) for the 5-day+ wet spells,

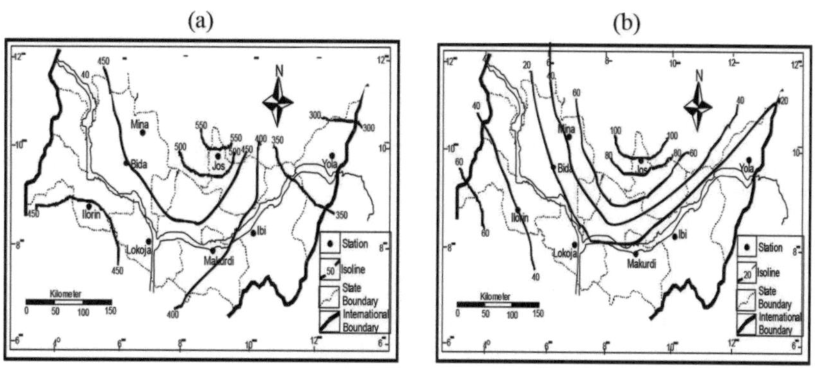

Figure 3.
Spatial distribution of total number of 2–4 days and 5-day+ wet spells in MBR.

Station	Duration of spell					
	2–4 days			5 days +		
	Total	Mean	CV	Total	Mean	CV
Bida	440	15	19.5%	20	1	113.7%
Ibi	381	13	27.9%	10	0	164.0%
Ilorin	459	15	17.3%	40	1	74.6%
Jos	557	19	18.3%	100	3	31.7%
Lokoja	430	14	27.1%	26	1	120.2%
Makurdi	422	14	23.0%	20	1	113.5%
Minna	490	16	21.2%	49	2	74.5%
Yola	329	11	34.0%	14	0	224.2%
Region	439	15	22.78%	35	1	192.50%
CV range			16.7%			192.5%

Table 3.
Descriptive statistics of wet spells in the Middle Belt of Nigeria (1981–2010).

respectively. For wet spells, only the annual frequencies of Jos, Minna and Ilorin were higher than the areally average region for both the 2–4- and 5-day+ wet spells. (**Table 3**).

4.2 Trend in dry and wet spells

4.2.1 Trend in dry spells

The correlation coefficients, which show the strength and significance of trend in the four categories of dry spells in the Middle Belt, are presented in **Table 4**. For the 2–4-day dry spell category, two areas within the Niger-Benue valley manifested significant opposite trends, which are Ibi, with a negative and significant trend, and Lokoja with a positive and significant trend. The other areas within the valley had positive but insignificant trends. The case of Bida is almost steady but increased slightly. The rest of the region, from the northwest to the north central plateau of Jos, and the south western parts (Ilorin), showed a negative but insignificant trend.

The trend of 5–7-day dry spell category showed that the Niger-Benue valley areas (Ibi, Makurdi and Yola) had a positive but insignificant trend except for Lokoja and the region where the trend was negative but insignificant. The rest of the region experienced an insignificant decreasing trend, except the north central plateau zone which manifested positive but insignificant. This means that there had only been slight changes in the occurrences of the medium dry spell category (5–7 days) in the MB region, from 1981 to 2010.

The 8–10 days dry spell category showed equal decreasing and increasing trends among the eight stations (**Table 4**). The Lokoja, Makurdi and Ibi-Yola, and the averaged region experienced a decreasing trend which, however, was only significant at Yola, whereas the northwest (Minna), north central (Jos) and south western parts (Ilorin) witnessed an increasing trend, but it was only significant in the south west. The directions of these trends are shown in **Table 4**.

The 11 days and above dry spell category decreased significantly over Ilorin, Yola and the averaged region, while the north central plateau (Jos) and the Niger-Benue

Station	Duration of spell			
	2–4 days	**5–7 days**	**8–10 days**	**11 days+**
Bida	−0.026	−0.080	0.278	0.024
Ibi	−0.309*	0.295	−0.170	0.153
Ilorin	−0.167	−0.299	0.306*	−0.306*
Jos	−0.131	0.123	0.119	−0.197
Lokoja	0.334*	−0.252	−0.079	−0.143
Makurdi	0.180	0.047	−0.273	0.073
Minna	−0.170	−0.263	0.290	0.080
Yola	0.101	0.050	−0.367*	−0.306*
Region	0.020	−0.146	−0.137	−0.388*

Station / **Duration of spell** are bold headers.

**Significant trend at 95% confidence level.*

Table 4.
Correlation coefficients of dry spells in the Middle Belt of Nigeria (1981–2010).

DOI: http://dx.doi.org/10.5772/intechopen.1003859

Station	Duration of spell	
	2–4 days	5 days+
Bida	0.147	0.160
Ibi	0.003	−0.150
Ilorin	0.088	0.229
Jos	−0.038	0.368*
Lokoja	0.191	0.117
Makurdi	0.189	0.224
Minna	−0.024	0.159
Yola	0.369*	0.338*
Region	0.212	0.451*

**Significant trend at 95% confidence level.*

Table 5.
Correlation coefficients of wet spells in MBR (1981–2010).

confluence area (Lokoja) experienced decreasing but insignificant trend. Elsewhere, the trend was positive but insignificant (**Table 4**). The correlation result (**Table 4**), which showed a diffused situation whereby the probability of increasing trend of annual frequency of dry spells (positive correlation), points to severe drought and decreasing trend of annual frequency of dry spell (negative correlation), and points to a mild drought in the region.

4.2.2 Trend in wet spells

The trend in annual frequency of the 2–4 days wet spells showed that only Jos and Minna have negative trends (**Table 5**). The rest areas of Yola, Lokoja, Makurdi and Ibi, and Ilorin, and the areally-averaged region had positive trends. However, only the positive trend of Yola was significant.

For the long-wet spell category (5 days and above), there was a generally positive trend in the Middle Belt region, except for Ibi. However, only the positive trends of Jos, Yola and the region were significant (**Table 5**).

5. Discussion

The orographic effect of the Jos Plateau has resulted not only in higher rainfall amounts but also increase in rain days and wet spells on the Plateau. The effect of the Jos Plateau, which is clearly discernible in the spatial pattern of dry and wet spells in the Middle Belt, is also documented in other rainfall characteristics such as on onset, cessation and duration of rainy season, daily rainfall organisation and intensity and annual rainfall [31–34]. The Jos Plateau accelerated convective activities given rise higher annual number of rain days recorded in the region [32]. The zonal spatial pattern of increase (decrease) from the southern to the northern part of the region may be related to the proximity to the rain-bearing southwest trade wind, which advects moisture from the Atlantic Ocean. A similar zonal pattern is observed in the spatial distribution patterns of eight rainfall characteristics in the Middle Belt region [31].

For the entire region, there are more increasing trends of wet spells than decreasing trends. This is significant to agriculture and water resources management, especially soil erosion and flood control and management. In addition, as rain spells are related to synoptic systems such as depressions, cyclonic systems, fronts and troughs [1, 26, 29], the increasing trends of wet spells in the Middle Belt region could be an indication of increased activities of synoptic systems, especially the African easterly wave (AEW) that favours rainfall occurrence in central Nigeria.

6. Conclusion

The annual frequencies of 2–4- and 5-day+ wet spells were higher in the southern part than the northern part of the Middle Belt region. Conversely, the annual frequencies of the four categories of dry spells showed higher values in the northern part of the region. However, there is a station-to-station variation in annual frequencies of both dry and wet spells with Jos having the highest frequencies, especially for the wet spells due to the orographic effect of the Jos Plateau with an elevation of about 1500 m.

The region experienced a generally increasing trend in wet spells compared to dry spells. An increase in the frequency of wet spells may potentially increase the risk of flooding in the region, especially along the banks of Rivers Niger and Benue. An increase in the frequency of dry spells may aggravate drought and desertification, especially in the northern parts of the region. Sustainable water and land management practices may mitigate the dual threat of drought and flooding in the Middle Belt region.

Acknowledgements

The support by the Cartographical Unit, Department of Geography, Benue State University, Makurdi in producing the isopleths maps is appreciated.

Conflict of interest

The authors declare no conflict of interest.

Author details

Bernard Tarza Tyubee[1*] and Michael Terver Iwan[2]

1 Faculty of Environmental Sciences, Department of Geography, Benue State University, Makurdi, Nigeria

2 Department of Geography, College of Natural and Applied Sciences, Kwararafa University, Wukari, Nigeria

*Address all correspondence to: btyubee@bsum.edu.ng

References

[1] Osei A, Quansah. Characterization of dry and wet spells and associated atmospheric dynamics at the Pra river catchment of Ghana, West Africa. Journal of Hydrology Regional Studies. 2022;**34**:1-29

[2] Ajayi AE, Olufayo AA. The sequence of wet and dry days at Ibadan and Onne (sub-humid zone of Nigeria). Nigerian Journal of Technology. 2002;**21**(1):38-44

[3] Deka S, Borah M, Kakaty SC. Statistical modelling of wet and dry spell frequencies over North-East India. Journal of Applied and Natural Science: 2010. 2010;**2**(1):42-47

[4] Sukla MK, Mangaraj AK, Sahoo LN, Sethy KM. A comparative study of three models for the distribution of wet and dry spells in the Mahanadi Delta. New York Science Journal. 2012;**5**(11): 54-61. Available from: http://www.sciencepub.net/newyork

[5] Mirmousavi SH. Examining the probable length in days of wet and dry spells in Khuzestan province. Quarterly Journal of the Hungarian Meteorological Service. 2015;**119**(4):475-491

[6] Fajana AL. Dry and Wet Spell Variability over Nigeria [Thesis]. Akure, Nigeria: Department of Meteorology and Climate Science, submitted to the School of Post Graduate Studies, Federal University of Technology; 2019

[7] Frei C, Davies HC, Gurtz J, Schär C. Climate dynamics and extreme precipitation and flood events in Central Europe. Integrated Assessment. 2000;**1**:281-299

[8] Palmer TN, Räisänen J. Quantifying the risk of extreme seasonal precipitation events in a changing climate. Nature. 2002;**415**(31):512-514

[9] Federal Government of Nigeria [FGN]. Nigeria Post Disaster Needs Assessment: A Report on the 2012 Floods. Federal Government of Nigeria [FGN]; 2013. pp. 15-56

[10] Ezeh CU, Obeta MC, Anyadike RNC. Variations in the sequences of daily rainfall across Nigeria. Arab Journal of Geosciences. 2016;**9**:681. DOI: 10.1007/s12517-016-2719-9

[11] Yapo ALM, Diawara A, Kouassi BK, et al. Projected changes in extreme precipitation intensity and dry spell length in Côte d'Ivoire under future climates. Theoretical and Applied Climatology. 2020;**140**:871-889. DOI: 10.1007/s00704-020-03124-4

[12] Anagnostopoulou C, Maheras P, Karacostas T, Vafiadis M. Spatial and temporal analysis of dry spells in Greece. Theoretical and Applied Climatology. 2003;**74**(1-2):77-91. DOI: 10.1007/s00704-002-0713-5

[13] Deni SM, Jamaludin S, Zawiah W, WZW WZ, Jemain AA. Tracing trends in the sequences of dry and wet days over peninsular Malaysia. Journal of Environmental Science and Technology. 2008;**1**:97-110. Available from: http://scialert.net/abstract/?doi=jest.2008.97.110

[14] Sivakumar MVK. Empirical analysis of dry spells for agricultural applications in West Africa. Journal of Climate. 1992;**5**:532-339. Available from: www.ametsoc.org [Accessed: April 7, 2016]

[15] Tyubee BT, Iwan MT. A Markov chain analysis of wet and dry spells

for agricultural crop planning in the middle belt region of Nigeria. Journal of Agriculture and Environmental Sciences. 2019;**8**(2):132-147

[16] Gobin A, Van de Vyver H. Spatio-temporal variability of dry and wet spells and their influence on crop yields. Agricultural and Forest Meteorology. 2021;**308-309**:108565, ISSN 0168-1923. DOI: 10.1016/j.agrformet.2021.108565

[17] Fall CMN, Lavaysse C, Kerdiles H, Drame MS, Roudier P, Gaye AT. Performance of dry and wet spells combined with remote sensing indicators for crop yield prediction in Senegal. Climate Risk Management. 2021;**33**:1-27

[18] Okonkwo NN, Balogun A, Omotosho J, Agele S. Critical Dry Spell Prediction in Rain-Fed Maize Crop Using Artificial Neural Network in Nigeria. London, UK: IntechOpen; 2021. DOI: 10.5772/intechopen.100627

[19] Adane BA, Hirpa BA, Lim C-H, Lee W-K. Spatial and temporal analysis of dry and wet spells in upper Awash river Basin, Ethiopia. Water. 2020;**12**:3051. DOI: 10.3390/w12113051

[20] Mathlouthi M, Lebdi F. Comprehensive study of the wet and dry spells and their extremes in the Mediterranean climate basin North Tunisia. SN Applied Sciences. 2021;**2021**:3850. DOI: 10.1007/s42452-021-04834-8 [Accessed: August 7, 2023]

[21] Barkotulla MB. Stochastic generation of the occurrence and amount of daily rainfall. Pakistan Journal of Statistics and Operation Research. 2010;**6**(1):61-73. Available from: http://www.pjsor.com/index.php/pjsor/article/view/122

[22] Fetih Y, Turker BO. Modeling dry and wet spells of Central Anatolia region of Turkey. In: Proceeding of the 58th World Statistical Congress, 2011, Dublin (Session CPS015). Int. Statistical Inst; 2011. pp. 4195-4199. Available from: http://2011.isiproceedings.org/papers/950965.pdf

[23] Mathugama CS, Peiris S. Spatial and temporal analysis of critical dry spells in Sri Lanka, impacts & responses. International Journal of Climate Change. 2012;**3**(3):71-87

[24] Shahraki N, Bakhtiari B, Ahmadi MM. Markov chain model for probability of dry, wet days and statistical analysis of daily rainfall in some climatic zone of Iran. Karnataka Journal of Agricultural Science. 2012;**21**(1):399-406

[25] Admasu W, Tadesse K, Yemenu F, Abdulkadir B. Markov chain analysis of dry, wet weeks and statistical analysis of weekly rainfall for agricultural planning at Dhera, Central Riff valley region of Ethiopia. African Journal of Agricultural Research. 2014;**9**:2205-2212

[26] Striem HL, Rosenan N. Rainspells as a climatological parameter used in analysis of rainfall in Jerusalem. Archiv für Meteorologie, Geophysik und Bioklimatologie, Serie B. 1973;**21**:25-42

[27] Pena M, Douglas MW. Characteristics of wet and dry spells over the pacific side of Central America during the rainy season. Monthly Weather Review. 2002;**130**:3054-3073

[28] Ratan R, Venugopal V. Wet and dry spell characteristics of tropical rainfall. Water Resources Research. 2013;**2013**(49):1-12. DOI: 10.1002/wrcr.2075

[29] Froidurot S, Diedhiou A. Characteristics of wet and dry spells in the West African monsoon system. Atmospheric Science Letters. 2017;**18**:125-131. Available from:

https://www.researchgate.net/publication/313425222_Characteristics_of_Wet_and_Dry_Spells_in_the_West_African_Monsoon_System

[30] Di Guiseppe E, Vanto O, Epifani C, Efpofito D. Analysis of dry and wet spells from 1870-2000 in four Italian sites. Geophysical Research Abstracts. 2005;**7**:7-12

[31] Tyubee BT. Fluctuations and trends in rainfall characteristics over the middle belt of Nigeria. Journal of the Nigerian Meteorological Society (NMS). 2004;**4**(1):63-72

[32] Tyubee TB. Spatial organisation of daily rainfall in the middle belt of Nigeria. The Benue Valley: Journal of Interdisciplinary Studies. 2005;**4**(1):69-79

[33] Tyubee TB. Persistence and periodicity in rainfall characteristics in the middle belt of Nigeria. Journal of Geography and Development. 2006;**1**(1):1-7

[34] Mage OJ, Tyubee BT. Temporal trends in daily rainfall intensity in a changing climate in the middle belt region of Nigeria. European Journal of Geography. 2017;**8**(4):103-117

[35] Hernaeza PF, Martin-Vide J. Regionalization of the probability of wet spells and rainfall persistence in the Basque Country (Northern Spain). International Journal of Climatology. 2011;**2011**(32):1909-1920

[36] Fischer BMC, Mul ML, Savenije HHG. Determining spatial variability of dry spells: A Markov-based method, applied to the Makanya catchment, Tanzania. Hydrology and Earth System Sciences. 2013;**17**:2161-2170

[37] Adefisan EA, Abatan AA. Agroclimatic zoning of Nigeria based on rainfall characteristics and index of drought proneness. Journal of Environment and Earth Science. 2015;**5**(12):115-127. Available from: www.iiste.org

[38] Omotosho JB, Balogun AA, Ogunjobi K. Predicting monthly and seasonal rainfall, onset and cessation of the rainy season, in West Africa, using only surface data. International Journal of Climatology. 2000;**20**:865-880

[39] Tarhule A, Woo M. Changes in rainfall characteristics in Northern Nigeria. International Journal of Climatology. 1998;**18**:1261-1271. Available from: http://rmgsc.cr.usgs.gov/outgoing/threshold_articles/Tarhule_Woo1998.pdf

[40] Bärring L, Holt T, Linderson M, Radziejewski M, Moriondo M, Palutikof JP. Defining dry/wet spells for point observations, observed area averages, and regional climate model gridboxes in Europe. Climate Research. 2006;**31**:35-49. Available from: https://lup.lub.lu.se/search/.../9b9e1419-1052-45a7-833d-2980a1d50a4e

[41] Simba FM, Seyitini L, Murwendo T, Mapurisa B, Chayangira J, Matete C, et al. Analysis of trends in dry spells during rainy seasons for Masvingo airport and Zaka stations in Masvingo province, Zimbabwe. Journal of Environment Research and Development. 2012;**7**(1):218-228

[42] Gitau W. Diagnosis and predictability of intraseasonal characteristics of wet and dry spells over equatorial East Africa [thesis]. Kenya: University of Nairobi; 2013. Available from: https://www.google.com/search?q=Diagnosis+and+predictability+of+intraseasonal+characteristics+of+wet+and+dry+spells+over+equatorial+east+Africa&ie=utf-8&oe=utf-8&client=firefox-b&gfe_rd=cr&ei=61C0V6QLx8jyB5eeh_AL

Section 2

Rain Water Quality Study

Chapter 3

Water Isotopologues Long-Term Continuous Rainfall Monitoring Contribution for Modeling Present Climate: Information Obtained from Different Time Scale Observations

Didier Gastmans, Vinicius dos Santos,
Zayra Christine Sátyro dos Santos and Vladimir Eliodoro Costa

Abstract

Rainfall isotopic composition has been continuously monitored at the central portion of the São Paulo state (Brazil) in different sampling time scales since 2013. The integration of different meteorological data, such as surface data from meteorological stations, HYSPLIT trajectories, reanalysis and ERA-interim data, has led to observed different conclusions based on the isotopic observation time scale. The amount effect in tropical areas is important for isotopic monthly data, explaining classical effects on monthly data, such as seasonality (high (low) isotopic composition during the dry (wet) period). Based on a daily scale, the interpretation is more complex, leading to controls on isotopic composition related to moisture source/transport and convective activity, as well as some local factors. Using microrain radar with GOES-16 imagery to identify the rainfall type, we were able to understand the cloud microphysics and sub-cloud processes responsible for rain isotope composition variation during the event. This combination of isotopic data may provide substantial subsidies and information for coupling isotopic data in GCMs. The incorporation of water isotopes into GCMs has enabled a more comprehensive evaluation of the water cycle and improvements in hydrometeorological simulations. This contribution has provided new insights into present, past, and future climate.

Keywords: stable isotopes, rainfall, Brazil, and rainfall isotopic long-term monitoring, GNIP

1. Introduction

The temporal and spatial monitoring of the atmosphere, ocean, and land surface has evolved rapidly in recent decades due to the technological revolution

[1, 2], developing the World Weather Watch Global Observing System (GOS) [3]. The GOS has contributed to the understanding of climate phenomena, generating quality data that allows atmospheric scientists to measure, estimate, model, and unravel the past and project the future of the Earth's climate. Among the range of knowledge about climate produced in recent decades, it has been highlighted that man has become a direct agent of meteorological processes, promoting climate change [4].

The perception of extreme and harmful events in our daily lives has become increasingly pronounced due to ongoing changes. It is possible to gauge the occurrence of these extreme weather events on a global scale through the maintenance of the GOS, as reported by the World Meteorological Organization (WMO) [5]. According to the WMO, between 1970 and 2019, there were more than 11,000 disasters attributed to climate and water hazards, which accounted for just over 2 million deaths and US$3.64 trillion in losses [5]. Assessments like this one become possible not only by improving the quality of meteorological data using technologies but by maintaining them over time, allowing comparison with information from the past.

Centuries-old surface weather stations (https://public.wmo.int/en/our-mandate/what-we-do/observations/centennial-observing-stations) are a clear example of the importance of long-term observations, constituting the backbone of global meteorological coverage, serving as a basis for climatology and weather forecasting studies, as they allow an incisive assessment of climate variability.

Another incisive example of long-term monitoring also encouraged by the WMO in partnership with the International Atomic Energy Agency (IAEA) was the creation of the Global Network of Isotopes in Precipitation (GNIP) (https://www.iaea.org/services/networks/gnip). The GNIP was created with the initial objective of monitoring the concentrations of tritium (^{3}H – radioactive isotope) in the atmosphere, produced by nuclear tests during the interwar period. Later, stable isotopes of rainwater, oxygen (^{18}O, ^{17}O, ^{16}O), and hydrogen (^{2}H, ^{1}H) also became part of the monitoring scope, which began in the 1960s. Presently, GNIP has a monthly, online, and free database of the isotopic composition of precipitation and monthly meteorological data (precipitation and temperature) from more than 1200 stations installed in ~100 countries around the world. This database has been widely used in studies on global climate, meteorology, ecology, and hydrology, contributing to verifying and improving climate models, and hydrological models, involving surface water and aquifers at different spatial scales.

In Brazil, 28 GNIP stations were installed and rain samples were collected between 1957 and 1990 [6]. Despite having different objectives, these stations generated results from the first isotopic studies in the national territory, with emphasis on studies carried out in the Amazon, with the aim of understanding the role of the forest in generating humidity and its influence on the local hydrological cycle. Unfortunately, this effort has been discontinued, generating a large gap in isotopic data, consequently on part of the understanding of the movement of water in several hydrographic basins and its relationship with atmospheric processes. Only in 2008, a new NGIP was installed in Brazil, in the city of Belo Horizonte, followed by Rio Claro (2013), both in the southeastern region of the country. The Geological Survey of Brazil (SBG-CPRM) and the National Water Agency (ANA), with the support of the IAEA, resumed the operation of a monthly monitoring network of the isotopic composition of rainfall in several regions of Brazil, contributing to subsidize several hydrogeological studies in the future.

There is no way to project the future without assessing the past, so this present gap in monitoring the isotopic composition of rainwater, rivers, and aquifers has impaired an integrated assessment of atmospheric and hydrological processes in Brazilian watersheds, especially in the context of climate changes. Despite climate change being named as global, it is in the local context of each river basin that the main changes in the hydrological cycle are effectively felt. Understanding how these local hydrological changes are related to atmospheric processes of different temporal and spatial scales requires the use of tools capable of promoting this integration, and hence the great differential and applicability of the use of stable isotopes of water [7, 8]. For this assessment to be able to indicate what changes have been taking place over the years, long-term monitoring is essential.

In addition to the importance of stable isotopes for understanding hydrological studies, the implementation of water isotopes in General Atmospheric Circulation Models gives confidence to climate projections, as it represents physical–chemical processes without the need for large parameterizations and is perhaps one of the most useful for understanding climate processes on a global scale, involving precipitation and atmospheric circulation regimes [9].

In this context, the development, evolution, and current status of monitoring the isotopic composition of rain at the Rio Claro GNIP station (code: 8374701), operating from February 2013 to the present date, will be presented, revealing its contribution to the understanding of the atmospheric processes of the present climate. During this collection period, several atmospheric phenomena occurred and contributed to the formation of rain and changes in Rio Claro weather. Among these phenomena, some examples, such as the influence of El Niño-Southern Oscillation events during different periods in its warm phase (2014–2016, 2018–2019) and cold phase (La Niña, 2016, 2016–2017, 2020–2021, 2021–2023) according to the ONI-INDEX, extreme events related to hailstorms [10], above-average daily volume (~100 mm/day) and decreased rainfall for a period of extreme drought [11].

The observation of these and other atmospheric rainfall systems and their relationship with the isotopic composition of water was only possible due to continuous and long-term monitoring carried out with the monthly sampling (a monthly composite sample collected on the first day of every month at 12UTC) and with daily collection (e.g., a sample collected between 12:00 UTC on the 9th and 12:00 UTC on the 10th) started in February 2014. Subsequently, seeking to expand the assessment of the isotopic composition of rain, a collection was carried out at high frequency (in minutes, also named intra-event) to assess isotopic evolution during the passage of an individual rainfall event.

Thus, combining different sampling time scales with meteorological data of different spatial and temporal resolutions, the main objective of this book chapter is to share how isotope monitoring at different collection scales contributes to the understanding of the meteorological processes related to rain. Two questions serve as a basis for determining the proposed objective: (i) How has monitoring evolved over time? (ii) What was revealed by each type of collection? In order to present this contribution, this chapter is subdivided into two sections: Section 2 presents the basic concepts about stable isotopes and their application in rain; Section 3 presents the comparison of the main interpretations of meteorological controls on isotopic composition between monthly, daily, and high-frequency collection scales; Section 4, discuss how and why the monitoring of the isotopic composition of rain was carried out at different collection scales; and finally the last section, where the main conclusions are presented.

2. Deciphering the history of water with stable isotopes

The word isotopes originated from the Greek: Iso "equal," topes "place," means that atoms occupy the same place, that is, the same position in the periodic table, being constituted of a different number of neutrons, therefore of different masses, resulting in molecules made up of "light" and "heavy" atoms. Considering the water molecule, called isotopologues, the stable isotopes of hydrogen (H) and oxygen (O) most used in atmospheric and hydrological sciences are ^1H (light) e deuterium (D) or ^2H (heavy), ^{16}O (light) e o ^{18}O (heavy), respectively [12, 13].

As the measurements of the amounts of these isotopes are not absolute, but represent the ratio between the least abundant isotope (heavy) over the most abundant (light), (for more details on the isotopic abundance see [12, 13]) the notation δ is used, expressed in parts per thousand (‰) and a reference standard for comparison (the δ of a sample collected in a given location is compared with this reference standard). The most used reference standard is the Vienna Standard Mean Ocean Water—VSMOW (Eq. (1)), which represents the average isotopic composition of ocean waters. Thus, positive values of δ indicate isotopic ratios that exceed the VSMOW and negative values of δ indicate ratios lower than the VSMOW [12].

$$\delta^{18}O\,ou\,\delta^{2}H=\left(\frac{\left(\frac{^{18}O(ou\ ^{2}H)}{^{16}O(ou\ ^{1}H)}\right)_{amostra}-\left(\frac{^{18}O(ou\ ^{2}H)}{^{16}O(ou\ ^{1}H)}\right)_{padrão}}{\left(\frac{^{18}O(ou\ ^{2}H)}{^{16}O(ou\ ^{1}H)}\right)_{padrão}}\right)\times 1000(‰) \qquad (1)$$

Throughout the hydrological cycle, from the process of evaporation of water molecules in the ocean (the main source of moisture in the world) to the formation of rain on the continents, where water molecules reach the earth's surface, recharging rivers, soils, and later aquifers, several water phase change processes modify the ratio of light and heavy isotopes in the water molecule. The physical-chemical process responsible for the partitioning of isotopes during water phase changes is called isotopic fractionation [12–15], forming distinct water molecules, which vary spatially and temporally, registering an unique isotopic signature, functioning as a "fingerprint" about the paths taken throughout this cycle, which makes it possible for scientists to tell the story of water.

In relation to precipitation, there are two types of fractionation that directly affect the isotopic composition of rainfall: (a) equilibrium fractionation, related to preferential exchanges that different substances have for a given isotope (occurs during the process of condensation inside clouds, formation of ice and rain droplets) [7, 16–18] and (b) kinetic or nonequilibrium fractionation, related to different rates of reaction between molecules (occurs in evaporation, isotopic exchange with surrounding vapor and reevaporation of drops in the rain) [17, 19].

In isotope studies of rain, relative terms, enriched and depleted, are generally used to denote whether the heavy isotope content is higher (rich in heavy isotopes) or lower (poor in heavy isotopes), respectively. In this sense, when water evaporates from the ocean, molecules with lighter isotopes ($^1H_2{}^{16}O$) tend to evaporate and form water vapor, depleted in heavy isotopes compared to the water that gave rise to it. Conversely, when it rains, molecules with heavier isotopes tend to precipitate ($^1HD^{16}O$ e or $^1H_2{}^{18}O$), forming an isotopic composition enriched in heavy isotopes [7, 12].

This difference between light and heavy water is mainly determined by the amount of fractionation that a given portion of steam suffered, generated rain and interacted with the surface, consequently with new processes of evaporation, evapotranspiration, and subsequent condensation, from its origin (source area) to the location where rainwater was collected. Thus, the isotopic composition of rain will be more depleted (loss of heavy isotopes along the path of the steam) the farther it is from its source of water [12, 15, 20]. This so-called Rayleigh distillation concept is essential for understanding the regional processes that affect the isotopic composition of rain, as it is related to the origin of steam, moisture transport, interaction with regional atmospheric circulation, and atmospheric systems, revealing the history of rainfall.

Locally, when rainfall falls on a collection point, a stage called post-condensation processes, it is also subject to two fractionation processes: (i) rainwater composition balances with the surrounding humidity and becomes enriched. This balance depends on droplet size and relative humidity, which at lower levels is less depleted than rainfall at the cloud base, so the isotopic composition of surface rain closely resembles that of surface moisture [21]; The larger the raindrop, the greater its falling velocity and the less exchange with the surrounding vapor, resulting in $\delta^{18}O$ depletion [9]; (ii) evaporation of the drops that fall on a layer of low humidity enriches the remaining rain, making the surface rainfall more enriched [22–26]. This process has been observed mainly in desert areas [15, 21, 27], although it is also seen in continental areas, mainly in light rains in a relatively dry atmosphere [20].

Thus, by combining regional and local fractionation processes, it is possible to determine the isotopic signature of rain, but what makes this identification possible is the way in which the isotopic composition of rain relates to geographic factors and climatic elements, i.e., the approach that is adopted, intends to used to interpret the history of rain. In this way, the sample collection scale determines the degree of interpretation that one intends to have on the isotopic composition of rain, so that regional processes can overlap local processes, and vice versa.

3. Evolution of sampling frequency scales and possibles interpretation of meteorological and isotopic processes

The dynamics between water phase changes and the fractionation process characterize the spatial and temporal variation of isotopic variability in different areas of the globe, since distances such as moisture source, latitude, climate, altitude, land use and land cover, and the acting atmospheric systems are entirely distinct. In this sense, isotopologues can be related to climate dynamics at different temporal and spatial scales, ranging from minutes to hundreds or thousands of years (climate variability), and spatially ranging from the micro-scale (less than 1 km), in atmospheric turbulence, to synoptic (>2000 km) involving large cloud assemblages (**Figure 1**) [13, 20, 28, 29].

Figure 1, summarizes the combination of spatial and temporal variation between sampling frequency and atmospheric systems, illustrating what spatial and temporal level a study can incorporate, resulting in different interpretations of how meteorological processes explain rainfall isotopic variability.

For the monthly sampling frequency, a composite sample is collected, in other words, the monthly isotopic signature represents the sum (of n events) or the rain-weighted average of the isotopic fractionation processes that occurred in all rainfall

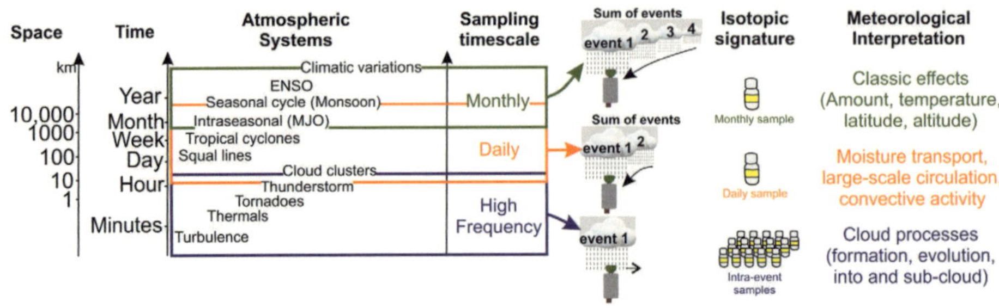

Figure 1.
Synthesis of the relationship between atmospheric systems, isotopic sampling scale, and main meteorological interpretations for understanding climate over time and space.

events in the interval of 1 month. With this isotopic signature, it is possible to interpret meteorological processes acting on a large spatial (>2000 km) and temporal (months and years) scale, such as intraseasonal, seasonal, ENSO (interannual) events, and even long-term climate variations, when isotopic monitoring occurs for at least 10 years [30]. Classical effects (such as temperature, latitude, altitude, continentality, and amount) are the main interpretations of the isotopic composition of rainfall.

The earlier research and interpretation were based on monthly samples, and it was developed as the background of concepts and interpretations of the isotopic composition of rainfall, using GNIP database. One of the main elements of the explanation of isotopic variability of meteoric waters is the Global Meteoric Water Line (GMWL), which represents a global linear relationship between $\delta^{18}O$-$\delta^{2}H$, defined by the equation: $\delta^2H = 8 * \delta^{18}O + 10$ [31]. Subsequently, using GNIP data, a new GMWL equation was computed ($\delta^2H = 8{,}17 * \delta^{18}O + 11{,}27$) [20], despite the Craig's equation is widely used. The GMWL is a reference for local studies in different climatic regions of the globe and compares to a Local Meteoric Water Line (LMWL), which reflects the average and local relationship between $\delta^{18}O$-$\delta^{2}H$. Since, for a given temperature range and isotopic composition, deviations from the LMWL occur by equilibrium ($\delta^{18}O$-$\delta^{2}H$ values around the GMWL) or kinetic processes (slopes of the line different from 8) [30, 32].

The LMWL provides an assessment of the spatial variation of rainfall isotopic composition by comparing isotopic data from different stations and identifying information on seasonal climatology [20, 31]. In addition to the comparison between rain waters, a LMWL serves as a reference to interpret the isotopic composition of other waters (soil, rivers, lakes, groundwater) and plants (stems and leaves), allowing the understanding of water movement and the interaction between these different compartments in a given watershed [7, 30, 33].

Another second-order parameter, which helps to explain the nonequilibrium (kinetic) processes of isotopic fractionation of H isotopes in relation to those of O, is the deuterium excess (*d*-excess), defined: $d = \delta^2H - 8 * \delta^{18}O$, represented by the value of 10 in GMWL [15] and deviations of LMWL. Its variation is usually related to temperature gradients in the regions where the vapor origin is located (most of it is oceanic), isotopic exchanges during vapor transport over continents ($d > 10‰$) [34, 35], local evaporation processes ($d < 10‰$), during the falling raindrops toward the surface [15, 21, 36] and in surface and groundwater studies (aquifer recharge) [6, 7, 37].

The application of these definitions has enabled a spectrum of interpretations of isotopic variability, characterized by the classic "effects" that have temperature dependence as an important element: In *latitude effect*, in the equatorial zone (warm) are more enriched isotopic values and in the poles (cold) more depleted isotopic values; *altitude effect*, occurs due to temperature decrease with increasing altitude, resulting in depletion of isotopic composition; *continentality* effect, associated with the transport of vapor and rainfall from the ocean to the continent, isotopic signatures in coastal areas are more enriched than continental areas (more depleted), explained by the Rayleigh distillation process; *temperature effect*, observed in seasonal variations in the northern hemisphere and at higher latitudes, where there is a large temperature range between summer (more enriched) and winter (more depleted). Differences in the rainfall regime over the year also characterize the amount of effect in tropical areas, defined as the negative linear relationship between $\delta^{18}O$ and the amount of rainfall, so the greater the amount of rainfall, the greater the depletion of the monthly isotopic composition due to successive condensation processes [12, 15, 20].

In accordance with the evolution of climate monitoring technologies associated with the improvement on isotopic determination with the implementation of new techniques such as Laser Absorption Spectroscopy has increased the capacity to determine the number of samples [38], enabling the expansion of isotopic monitoring to daily sampling and increased studies of individual rainfall events with high-frequency sampling.

For the daily sampling scale, a single daily isotopic signature represents the sum of one or more events collected during 1 day or the daily rainfall-weighted average of the isotopic fractionation processes that occurred in these rainfall events over the interval of 1 day. Using the daily isotopic signature is possible to interpret the meteorological processes that operate on a large spatial and temporal scale, as demonstrated by the monthly data from continuous long-term monitoring. However, the daily scale decreases the mixing and overlapping effect between rain events and consequently the different types of isotopic fractionations that occurred during the rain formation, enabling a better resolution to understand the isotopic variability of the different types of weather, identifying usual events from extreme events. For this reason, the daily scale improves the analysis of the interannual variations and seasonal cycles, as well as the action of atmospheric systems between 1000 km and 10 km, in the range of days to a week, such as tropical cyclones, squall lines, and cloud clusters (**Figure 1**). The daily isotopic composition of rainfall includes evaluations of the moisture origin/transport related to large-scale circulation systems, different types of rainfall, and the influence of convective activity. The daily isotopic composition of rainfall is also related to the classical effects that are better evaluated on a daily scale.

One of the main differences between the isotopic composition of monthly and daily rainfall was observed in the assessment of the amount effect in continental tropical areas. Strong correlations between monthly $\delta^{18}O$-rainfall decrease at the daily scale, due to several factors still under investigation. One of the best hypotheses explaining why daily and short-term isotopic variations do not correspond to rainfall amount, suggested the importance of convective processes in modulating the isotopic content of rainfall, reflecting the integrated history of convective activity over 4 days [36]. In organized convective systems, reevaporation processes of raindrops in mesoscale subsidence updrafts form a low-level depleted vapor in the atmosphere, feeding successive convective systems in that interval of integrated convective activity

[36]. Daily isotopic data provide an understanding of the evolution of this convective history, so the isotopic composition is not only related to its final rainfall amount, offering great proxy information for understanding how convection transforms water vapor into rain, after all this mechanism is essential in the energy balance and distribution of water across the globe.

At the high-frequency sampling scale, several samples are collected, hence the variation of several isotopic signatures over the evolution of an individual rain event is observed, forming an isotopic trend of variation, always mentioned in high-frequency studies [17, 21–23]. Evaluating the isotopic trend provides an interpretation of meteorological processes that operate at smaller spatial (>100 km) and temporal (hours to minutes) scales, such as cloud clusters, thunderstorms, tornadoes, and local rainfall (**Figure 1**). Using intra-event is possible to evaluate in detail several components, which involve the path, structure and evolution of the storm, the atmospheric system that originated it, changes in the air mass, altitude at which rain is produced (condensation level), type of rain, intensity of rainfall, and local and microphysical processes that occur inside and below the clouds, such as diffusive exchanges between low-level vapor and raindrops (main the evaporation of raindrops), and the relationship with surface meteorological data [17, 20, 21].

The application of these concepts are shown in the next chapter, based on the main results obtained during the 9-year monitoring of the isotopic composition of rainfall in Rio Claro, at different collection scales.

4. Monitoring the isotopic composition of rainfall in Rio Claro

Due to the need to understand how atmospheric processes control the formation and variability of rainfall, which is the main input in aquifer water recharge, rainfall stable isotope sampling stations were installed in several localities (Rio Claro, Brotas, and Araraquara), located in the central-southern portion of São Paulo state. This region is one of the recharge areas of the Guarani Aquifer System (SAG). In this context, the Rio Claro station was affiliated with GNIP, latitude: −22.39°S, longitude: −47.54°W and elevation of 670 m.a.g.l, starting with a monthly collection of stable isotopes of rainfall, sent to the IAEA laboratory in Vienna, along with rainfall and precipitation data.

In order to expand the monitoring of the isotopic composition of rainfall and understand the role of Amazon moisture in the formation of rainfall in the south-central portion of the state of São Paulo, southeastern Brazil, daily samples started to be collected in February 2014. Results, differences, and interpretations of meteorological controls on the isotopic composition of monthly and daily rainfall were discussed in Section 4.1.

The high-frequency rain sampling (details in Section 4.2) was important to understand the isotopic variability during the passage and evolution of different types of rainfall, convective and stratiform, since monthly and daily assessments did not help to understand how these rainfall types control the isotopic composition in Rio Claro.

The Rio Claro region has an average annual rainfall of around 1500 mm, characterized by two distinct seasons. The first season is a rainy and warm spring-summer period occurring between October and March. The second season, which is cooler and less rainy, occurs in the autumn-winter period between April and September. Among the primary regional weather systems, the Cold Fronts (polar air masses), active throughout the year. The South Atlantic Convergence Zone (SACZ) is prevalent in summer and the South Atlantic Subtropical High (SASH), which inhibits rain formation during winter and contributes to the transport of moisture from the Atlantic Ocean to the continent in the other seasons of the year. The Atlantic Ocean serves as the primary moisture source in Brazil, with additional contribution from the Amazon rainforest's evapotranspiration. The moisture from the Amazon is crucial to the formation of rainfall. It is transported across South America by low-level jets during the rainy season [39–42].

4.1 Monthly and daily

Between February 2013 and December 2022, 109 monthly samples were collected in Rio Claro station, resulting in a variation from −12.65‰ to 3.07‰, arithmetic mean and standard deviation of −4.73 ± 3.27‰ for $\delta^{18}O$, and from −94.90‰ to 23.70‰ (−23.46 ± 4.22‰) for $\delta^{2}H$, and from −0.86‰ to 24.12‰ (14.39 ± 4.22‰) for *d*-excess. Considering the mean $\delta^{18}O$ value, 56% of the samples were isotopic composition higher-4.73‰, while the remaining 44% can be considered depleted. For *d*-excess, lower values than <10‰ were observed in only 14% of the dataset, while 86% of these values were greater than >10%. These results characterize the dispersion of the isotopic composition values and help to visualize an overview of this variation, with slight distribution between enriched and depleted values, and lower influence of local processes (kinetic fractionation) that decrease *d*-excess values.

The $\delta^{18}O$ and $\delta^{2}H$ values were aligned around GMWL, characterized by the e Monthly LMWL ($\delta^2H = 8.12 + \delta^{18}O * 15.24$) with intercept close to the GMWL value (8) and higher slope (10) (**Figure 2A**), indicating a predominance of processes related to the continental moisture recycling, which explain the higher *d*-excess values observed.

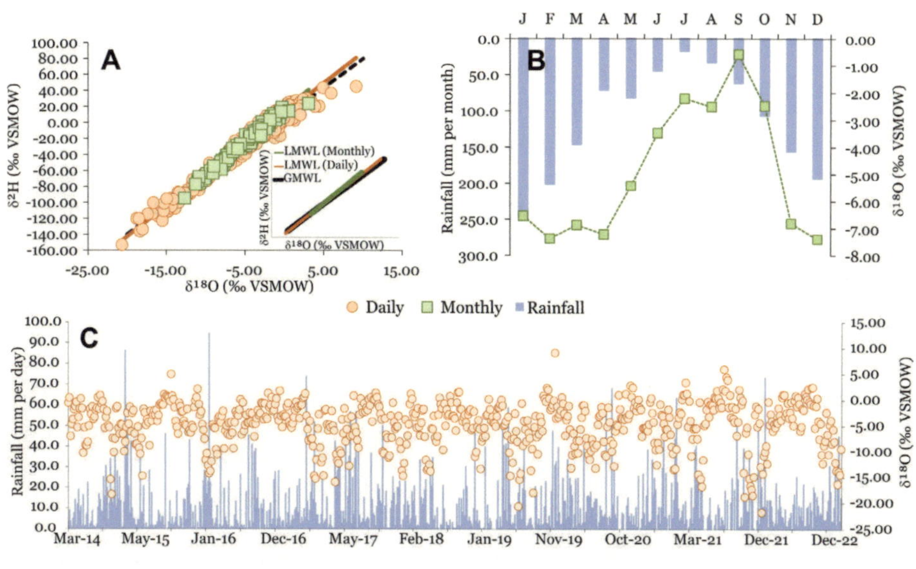

Figure 2.
Overview of the isotopic composition variation in Rio Claro station. (A) Classic $\delta^{18}O$-$\delta^{2}H$ relationship by monthly (2013–2022) and daily (2014–2022) dataset, global meteoric water line (GMWL) and monthly (green)/daily (orange) local meteoric water lines, (B) monthly average isotopic composition weighted by the amount of rainfall and monthly mean of rainfall, and (C) daily isotopic composition and rainfall.

One of the common practices in isotopic hydrology studies is the use of the average isotopic composition weighted by the amount of rainfall. This weighted average can be used for a seasonal, annual comparison, or for the full data set. In **Figure 2B**, the bars in blue are the monthly averages over the monitoring period and the points and lines in green are the average weighted by the amount of rainfall in each month ($\delta^{18}O_{wgd} = \sum (\delta^{18}O_{month\text{-}i} * P_{month\text{-}i}) / \sum P_{month\text{-}i}$). A clear seasonal distribution was observed, characterized by depleted values ($\delta^{18}O < -6.54‰$) during the rainy period (October–March) and enriched values ($\delta^{18}O > -5.43‰$) in the less rainy months (April–September). December ($\delta^{18}O$ -7.39‰) and February ($\delta^{18}O$ -7.38‰) were the most depleted months, while September was the most enriched ($\delta^{18}O$ -0.59‰) (**Figure 2B**).

The weighted averages presented more depleted values for $\delta^{18}O$ (−6.69‰) and δ^2H (−24.12‰) compared to the arithmetic mean, and lower for *d*-excess (13.14‰), illustrating that the amount of rainfall is more relevant for isotopes than *d*-excess. The inverse relationship between rainfall amount and monthly isotopic composition characterizes the amount effect, of strong negative and significant (p-valor <0.0001) correlations, r = −0.58 ($\delta^{18}O$) and – 0.60 (δ^2H). In contrast, the correlations between monthly temperature and isotopic composition was weak, r = −0.34 ($\delta^{18}O$) and – 0.37 (δ^2H), despite the significance ($p < 0.0002$).

The moisture-recycled transport by air masses (LMWL) and local amount effect controlled the monthly isotopic composition of rainfall by the condensation-related mechanism (Rayleigh distillation). The seasonal variations in isotopic compositing were distinct due to the different sources of moisture, transport, available moisture, and the performance of different atmospheric systems [38]. However, the dynamics of these main synoptic features change from day to day, since the monthly isotopic composition overlaps the observed variability, making the evaluation very focused on months of high and low rainfall [38].

The daily isotopic composition (represented as orange dots in **Figure 2A** and **C**), was evaluated based on 674 samples collected between February 2014 and December 2022. Despite the daily values varied aligned to monthly data, greater range of values were observed for $\delta^{18}O$, δ^2H and *d*-excess, −21.74‰ a 4.89‰ (−4.77 ± 4.34‰), −158.50‰ a 43.40‰ (−25.59 ± 34.45‰) e – 1.81‰ a 32.51‰ (12.62 ± 5.29‰), respectively. As observed in the monthly analysis, 57% of samples were more enriched (43% more depleted) in relation to the mean arithmetic $\delta^{18}O$ values, which is also very similar to the monthly arithmetic mean. Values of *d*-excess lower than <10‰ were observed in 27% of the dataset, while 73% of these values were greater than >10‰. These results reinforce the good distribution between enriched and depleted rainfall, with a greater influence of local processes (kinetic fractionation that decreasing *d*-excess) being observed with daily data in relation to monthly data.

This influence of the kinetic fractionation is characterized by the daily LMWL ($\delta^2H = 7.83 + \delta^{18}O * 11.84$), of a lower intercept and slope compared to the monthly LMWL, despite the close values in relation to GMWL. Continental moisture recycling processes also influence the isotopic composition of daily rainfall, since higher *d*-excess values are observed in most samples analyzed.

A clear seasonal variation in "V-shaped" (**Figure 2C**, black arrows) was observed for the daily isotopic composition of the rainfall. Enriched $\delta^{18}O$ values were predominant in the dry period, from April to September ($\delta^{18}O > -0.40‰$ and daily average rainfall of 7.2 mm/day), while depleted $\delta^{18}O$ values ($\delta^{18}O < -9.0‰$ and daily average rainfall of 19.8 mm/day) during the rainy season, between November and February (black circles highlighted in **Figure 2C**). The mean and standard deviation of daily rainfall during the complete monitoring period was 12.39 ± 13.62 mm/day.

The daily weighted average isotopic composition for the entire dataset was −6.15‰, −35.52‰ e 13.67‰ for $\delta^{18}O$, $\delta^{2}H$ and *d*-excess, respectively. Despite the weighting by quantify decreasing the values of $\delta^{18}O$, $\delta^{2}H$ and increasing those of *d*-excess in relation to the daily arithmetic mean, the amount effect was not observed for the daily scale. The weak and nonsignificant (p-valor >0.05) Spearman correlations were observed between daily rainfall and $\delta^{18}O$ (r – 0.24) and $\delta^{2}H$ (r – 0.22). In addition, the same occurred for daily temperature correlations of $\delta^{18}O$ (r – 0.10) and $\delta^{2}H$ (r – 0.13).

The role of the moisture source/transport mentioned in the evaluation of the monthly data became even more relevant on the daily analysis. Since, it is possible to observe the change in the transport of moisture every day, improving the understanding of the formation of rainfall in Rio Claro, from the origin to the successive condensation processes that occur along the history of rain and its relationship with the atmospheric systems over the course of the days. One of the most used tools in rain isotope studies is the Hybrid Single-Particle Lagrangian Integrated Trajectory (HYSPLIT) model, a mathematical system that calculates trajectories and simulates the dispersion and deposition of particles in the atmosphere (http://ready.arl.noaa.gov/HYSPLIT_traj.php) [43]. Meteorological parameters (e.g., rainfall rate (mm. h^{-1}), temperature (K), humidity (%), and trajectory height (meters) from different databases can be used to compose the meteorological information of the trajectory. Trajectories can be determined in backward mode (refers to pastime trajectories) or forward (future time trajectories), calculated by time determined by the user, as well as the coordinates and initial altitude.

In Rio Claro, back-trajectories were determined in different studies [44–46]. The Atlantic Ocean is the main moisture source, followed by the Amazon Forest and, in lower events from South Brazil. Between these source regions and Rio Claro station, along pathways of moisture interact with regional circulation, atmospheric systems, and convective activity, resulting in the moisture recycling process, illustrated by the monthly and daily LMWL. This mechanism occurs in different conditions between the seasons of the year, characterizing the observed seasonal variability.

During the rainy season, the moisture from the Amazon interacts with the pressure gradients formed by the increase in temperature, favoring the formation of the SACZ, resulting in more depleted isotopic values observed on the monthly and daily scale. In opposite, during winter, the moisture from the Amazon decreases, being more associated with the Atlantic Ocean, hence rainfall occurs when the FF is strong enough to overcome the circulation of a high-pressure system, such as the South Atlantic Subtropical High (SASH), resulting in a more enriched isotopic composition [44, 45, 47].

The comparison of the isotopic composition rainfall in Rio Claro, with other GNIP stations was also carried out, with the objective of spatially extending the Rio Claro analysis, confirming the role of these mentioned regional processes. This influence of available moisture and different moisture transport conditions was also observed for Belo Horizonte, Brasília, and Rio de Janeiro stations [47]. In this work, the seasonal, continental, and amount effects on the monthly isotopic composition of rainfall in all locations were confirmed. For the daily isotopic composition, one of the main meteorological controls on the variability in $\delta^{18}O$ and *d*-excess values was the strong convective activity (the deeper and more organized the convection, the greater depletion in the isotopic composition) during the summer. Meteorological parameters by reanalysis, such as Precipitable Water (kg m^{-2}), Outgoing Longwave Radiation (OLR - $W.m^{-2}$), 500 hPa vertical velocity field or Omega (Pa s^{-1}) were used to indicate the convective activity, and they were associated with daily isotopic composition.

A comparison of the isotopic composition of rainfall in different El Niño-Southern Oscillation (ENOS) events, 1997–1998 (ENOS 1) and 2014–2016 (ENOS 2), were carried out. The daily isotopic data from Rio Claro, Bragança Paulista, Campinas, Piracicaba e Santa Maria da Serra in the central-east portion of São Paulo state were used [48]. The same seasonal effect was identified, and mainly the influence of different available moisture conditions, which was higher in the dry season of ENSO 1, generating impoverishment in the isotopic composition of rainfall ($\delta^{18}O$ = <−4.60‰) compared to the dry season of ENSO 2, whose available moisture was lower and, consequently, the most enriched rainfall ($\delta^{18}O$ = <−2.80‰).

In addition, statistical tests presented in previous studies [45, 46] confirmed the influence of regional parameters on isotopic variability. In these studies, linear regression models were applied with significant results, explaining part of the isotopic variability, mainly values around $\delta^{18}O$ = −4.0‰ e − 5.0‰. The regression models, resulting in the need to investigate these rainfall events, did not explain strong negative and positive $\delta^{18}O$ values (**Figure 2C**).

Finally, the isotopic composition of daily rainfall was also associated with the distinction of different types of rain, convective (high $\delta^{18}O$ values) and stratiform (low $\delta^{18}O$ values) [49], relationship that has been widely investigated in different parts of the world. For Rio Claro, no good correlation was observed between $\delta^{18}O$ and convective and stratiform rainfall from two different databases, classification from the Global Precipitation Measurement (GPM) [50] and the ERA-interim (convective and large-scale precipitation converted in mm) provided by the European Center for Medium-Range Weather Forecasts (ECMRWF) [47].

4.2 High-frequency

For understanding the evolution of strong depleted and enriched rainfall events observed on a daily scale, identifying, and classifying rainfall types (convective and stratiform) and local processes related to the falling raindrops in Rio Claro, high-frequency collection was implemented.

A total of 312 samples were collected between 5, 10, and 30 minutes, from the beginning to the end of 18 rainfall events, between September 2019 and February 2021. As the collection was carried out manually, it was very difficult to be in the university facilities since the beginning of the rain. Therefore, the collection of events was performed randomly. Even so, in all seasons of the year, some intra-event was collected, covering a diverse range of atmospheric systems (frontal systems, prefrontal and postfrontal atmospheric instability, atmospheric instability thermal atmospheric during the summer, trough, and SACZ) and rainfall types (convective, stratiform, mixed (mixture between convective-stratiform during the same event), and localized rain) [51].

The isotopic composition of intra-event was combined with the meteorological data of high temporal resolution (minutes per hour), such as surface data (rainfall ($mm.min^{-1}$), temperature (°C), relative air humidity (%) and pressure (kPa), 1-minute interval) from meteorological automatic station (METER - Em50); vertical profile of the atmosphere over the collection point by micro rain radar (METEK MRR-2 operates at a frequency of 24.230 GHz with a modulation of 0.5–15 MHz) provide the reflectivity (Zc - dBZ), fall velocity (w – $m.s^{-1}$), rainfall rate ($mm.min^{-1}$) and liquid water content ($g.m^{-3}$), in 1-minute interval and height resolutions in a range bin of 31 measurement heights; ERA-5 vertical column of (humidity (%), temperature (°C), liquid water ($kg.m^{-3}$) and ice ($kg.m^{-3}$) content; imageries from the GOES-16 satellite (to identify convective nuclei and monitoring the formation and evolution of clouds with the brightness temperature (°C) of the images); determination of trajectories by the HYSPLIT model and regional data from ERA-5, easterly vapor flux ($kg.m^{-3}$), latent heat flux ($W.m^{-2}$) and OLR ($W.m^{-2}$) [51].

Rainfall types were classified employing a micro rain radar, GOES-16 imagery, and surface meteorological station. Radar images depicted distinct vertical structures for rainfall. Convective rainfall displayed a vertical structure, whereas the stratiform exhibited a horizontal structure. The horizontal structure's feature is the melting layer (or bright band in radar images) which can be quantified by a difference of approximately 4 dBZ in radar reflectivity between radar-measured heights. The identification of the melting layer determines the incidence of stratiform rainfall, and this analysis was conducted for all rainfall events evaluated in this study. The vertical structure is inadequate to confirm the occurrence of convective rainfall. The GOES-16 image was used to identify convective nuclei by analyzing a set of 40 pixels over Rio Claro with a brightness temperature lower than −38°C. The rainfall intensity was computed considering at least 10 mm per hour to determine convective rainfall. During the event, the mixed rainfall was a combination of convective and stratiform rainfall, while local rainfall was defined by the absence of a melting layer and convective nuclei [51].

This isotopic and meteorological dataset on a short temporal scale of minutes resulted in interesting explanations about the evolution of rainfall in Rio Claro, indicating different meteorological controls for convective, stratiform, mixed, and local rainfall.

Figure 3 illustrates an intra-event classified as mixed rainfall. This event occurred during the summer and was formed by predominant moisture from the Amazon Forest, according to HYSPLIT trajectories, where over the state of São Paulo it interacted with the passage of a cold front, characterizing a frontal system (warm mass from Amazon and cold mass from the polar) that produced cloud systems observed in GOES-16 image (**Figure 3A**). These systems generated rainfall over Rio Claro station, starting at 12:48 pm until 2:28 pm (local time, -3UTC).

In the first part of this event, the convective fraction was observed, characterized by a convective core in the GOES-16 image with a brightness temperature < −38°C in at least 40 pixels around the collection point (**Figure 3A**), structure vertical of the fall velocity by micro rain radar, without the presence of the melting layer (**Figure 3B**). In the second part of this event, a change in the vertical structure of the fall velocity was observed, with the presence of the melting layer (the values of w and reflectivity increase when this layer occurs), characterizing the stratiform fraction of this rain event (**Figure 3B**).

The isotopic composition of the rainfall responded to the evolution of the mixed event, with greater variation in $\delta^{18}O$ values (−10.29‰ ~ −11.61‰) during the convective phase and constant $\delta^{18}O$ values in the stratiform phase (−10.07‰ ~ −10.97‰) (**Figure 3C**). For *d*-excess the variation was inverse to the variation of $\delta^{18}O$. Lower d-excess values were observed during the transition from convective to stratiform phase (7.06‰) and at the end part of event (5.90‰ e 4.30‰) (**Figure 3C**). Relating the $\delta^{18}O$ values with rainfall rate, temperature, and relative humidity, it was observed that the lowest $\delta^{18}O$ value occurred in the interval of greatest rainfall intensity (**Figure 3C**), while temperature and humidity accompanied the small variation in $\delta^{18}O$ values during the entire period of the event (**Figure 3D**).

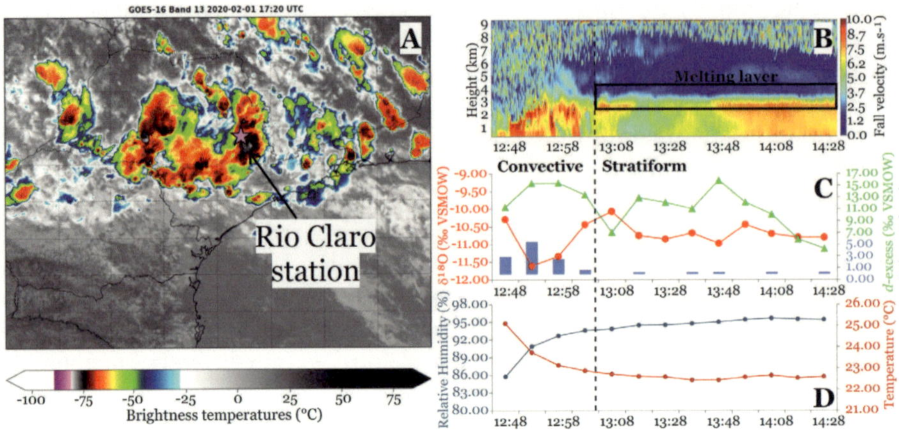

Figure 3.
Summary of local information used in the intra-event analysis. (A) GOES-16 image over the Rio Claro station, included convective nuclei around the station and in São Paulo state, (B) vertical profile of fall velocity (w) of micro rain radar for convective (vertical structure) and stratiform part (horizontal structure marked by the melting layer), (C) 18O (green), d-excess (red) and rain rates (blue), and (D) relative humidity (dark blue) and temperature (dark red). In B, C, and D, the vertical dot line separates the convective and stratiform parts of the event.

The event in **Figure 3** illustrated how the change in the rainfall types shifted the $\delta^{18}O$ and *d*-excess shape of variation, modulating the kinds of fractionation, in equilibrium (lower values of $\delta^{18}O$ in the highest rainfall intensity) and or kinetic (lower values of d-excess due to a change in the vertical structure of the rain, reflecting lower rain rates, hence the isotopic exchange between raindrops and low-level vapor) during the stratiform phase [51].

A comparison between intra-events of stratiform rainfall indicated that the isotopic composition represents the life cycle of the rain system. During the passage of the stratiform cloud over Rio Claro was observed that the rainfall events in the development phase were enriched in relation to events of the mature and dissipating phase (very depleted isotopic composition) [51].

High-frequency sampling not only provides an assessment of intra-event isotopic composition, but it can also facilitate an understanding of processes occurring over the course of a day, providing more than one isotopic signature. Thus, convective rainfall events reflected diurnal differences in isotopic composition, producing strong negative $\delta^{18}O$ values and higher d-excess during the night due to higher humidity conditions between the cloud base and the surface in relation to the daytime conditions, $\delta^{18}O$ less negative and smaller d-excess.

5. Conclusions

The continuous monitoring of the isotopic composition of rainfall in the GNIP Rio Claro station, Brazil, produced an interpretation of the present climate in the central-southern portion of São Paulo state, on different sampling scale. Monthly, daily, and high-frequency sampling scales were used like the vision instrument to see the dynamic between atmospheric systems linked to the rainfall and the water stable isotopes.

The monthly scale generated an overview of the classic effects (seasonal, rainfall amount, and continentality) in the isotopic composition of rainfall. These interpretations and LMWL are relevant references of isotopic signatures to relate the regional atmospheric dynamics with the local hydrological in the watersheds localized in the Rio Claro region.

The daily isotopic composition offered a more detailed interpretation viewfinder, characterized by monitoring the successive kinds of rainy weather (from synoptic to local spatial scales) day by day (from day to a week), resulting in temporal isotopic observations with clear seasonal patterns linked to stronger convective activity and moisture transport from Amazon during spring-summer (low isotopic composition) and cold fronts activity combined with moisture transport from Atlantic ocean during autumn-winter (high isotopic composition).

For high-frequency or intra-event scale, the interpretation view is enlarged like a magnifying glass zoom, about the formation and evolution of a rain system locally. The rainfall types and their influences on the isotopic composition were improved, clarifying the strong $\delta^{18}O$-$\delta^{2}H$ values and revealing new meteorological approaches like the diurnal isotopic composition differences linked to the day-night convection, and life cycle of stratiform rainfall.

Therefore, the history of the isotopic composition of rainfall in Rio Claro is characterized by the Atlantic Ocean and Amazon Forest moisture origins, transported over the long continental pathways resulting in moisture recycled. This mechanism was observed in all sampling scales, with a major emphasis on the daily scale. During these pathways, the moisture interacted with regional circulation, atmospheric systems, air masses, and convective activity removing the heavy isotopes in rainfall and forming the subsequently depleted vapor and next rainfall until arrived at the collection point. Locally, the convective and stratiform rainfall change the variations of $\delta^{18}O$ and *d*-excess values that reflect their different formation, including the microphysical processes into cloud and the processes in the below cloud base during falling raindrops (vertical structure of rainfall) at the surface (meteorological data).

Our results provide meteorological observation data and key insights for isotopic interpretation in tropical areas. Based on the acquired knowledge, it will be possible to use isotopic information and establish its relationship with the climate and the occurrence of extreme rainfall events to improve the understanding of past atmospheric processes (recorded in groundwater, glaciers, and spelotems), recent (plants, river water, rain) and future (indicative of water scarcity and increase in extreme events). By studying past events, it is possible to improve our management of water usage and availability considering climate change scenarios. Furthermore, implementing the use of isotopic methods in researching lesser-known hydrological systems can provide solutions to the challenges faced in managing water and climate resources, specifically those impacting tropical regions.

Acknowledgements

The research that produced the isotopic monitoring program in monthly, daily, and high-frequency scales were funded by grants from the São Paulo Research Foundation (FAPESP) under Processes 2015/15749-2 and 2018/06666-4, and by the International Atomic Energy Agency grant BRA-17984 under the initiative CRP-F31004 “Stable isotopes in precipitation and paleoclimatic archives in tropical areas to improve regional hydrological and climatic impact models,” and BRA-23531 under the initiative CRP-F31006 “Isotope Variability of Rain for Assessing Climate Change Impacts.” VS thanks FAPESP for the scholarship provided under the Processes

2013/06704-0, 2016/18735-5, 2019/03467-3 and 2021/10538-4. This study also was financed in part by the Coordination of Superior Level Staff Improvement (CAPES).

Conflict of interest

The authors declare no conflict of interest.

Author details

Didier Gastmans[1*], Vinicius dos Santos[1], Zayra Christine Sátyro dos Santos[2] and Vladimir Eliodoro Costa[3]

1 São Paulo State University Júlio de Mesquita Filho, Environmental Studies Center, Rio Claro, SP, Brazil

2 State University Júlio de Mesquita Filho, Environmental Studies Center, Rio Claro, SP, Brazil

3 São Paulo State University Júlio de Mesquita Filho – Center of Stable Isotopes Prof. Dr. Carlos Ducatti, Botucatu, SP, Brazil

*Address all correspondence to: didier.gastmans@unesp.br

References

[1] Bauer P, Thorpe A, Brunet G. The quiet revolution of numerical weather prediction. Nature. 2015;**525**:47-55. DOI: 10.1038/nature14956

[2] Trenberth KE, Koike T, Onogi K. Progress and prospects for reanalysis for weather and climate. Eos Transactions AGU. 2008;**89**(26):234-235. DOI: 10.1029/2008EO260002

[3] Thépaut J-N, Andersson E. The global observing system. In: Lahoz W, Khattatov B, Ménard R, editors. Data Assimilation: Making Sense of Observations. Vol. 1. Berlin, Heidelberg: Springer; 2010. pp. 263-281. DOI: 10.1007/978-3-540-74703-1

[4] IPCC. 2021: Summary for policymakers. In: Masson-Delmotte V, Zhai P, Pirani A, Connors SL, Péan C, Berger S, et al, editors. Climate Change 2021: The Physical Science Basis. Contribution of Working Group I to the Sixth Assessment Report of the Intergovernmental Panel on Climate Change. In Press. Available from: https://www.ipcc.ch/report/ar6/wg1/

[5] World Meteorological Organization. In: Douris J, Kim G, editors. WMO Atlas of Mortality and Economic Losses from Weather, Climate and Water Extremes (1970-2019). WMO-n°1267. Geneva, Switzerland; 2021. Available from: https://library.wmo.int/idurl/4/57564

[6] Gastmans D, Garpelli LN, dos Santos V, de Lima C, Quaggio CS, Santarosa LV, et al. Contribuição dos isótopos estáveis da água (H e O) no conhecimento dos aquíferos brasileiros: Estado da arte e perspectivas futuras. Derbyana. 2021;**42**:1-47. DOI: 10.14295/derb.v42.734

[7] Jasechko S. Global isotope hydrogeology—Review. Reviews of Geophysics. 2019;**57**:835-965. DOI: 10.1029/2018RG000627

[8] Mook WG. Environmental isotopes in the hydrological cycle. Principles and Applications. Technical documents in Hydrology, n°39. Vol. 1. Paris: UNESCO; 2000

[9] Xi X. A review of water isotopes in atmospheric general circulation models: Recent advances and future prospects. International Journal of Atmospheric Sciences. 2014;**2014**:1-16. DOI: 10.1155/2014/250920

[10] Maia DC, Fontão PAB, Souza LB, Christofoletti ALH, de Azevedo TS. Hail rain and its geographical repercussions in the countryside of São Paulo (SP) state. Mercator. 2019;**18**:1-16. DOI: 10.4215/rm2019.e18006

[11] Coelho CAS, Cardoso DHF, Firpo MAF. Precipitation diagnostics of an exceptionally dry event in São Paulo, Brazil. Theoretical and Applied Climatology. 2016;**125**:769-784. DOI: 10.1007/s00704-015-1540-9

[12] Clark I, Fritz P. Environmental Isotopes in Hydrogeology. Boca Raton: CRC Press; 342AD

[13] Mook WG. Isótopos ambientales en el ciclo hidrológico: principios y aplicaciones. Madrid: Instituto Geológico y Minero de España; 2002

[14] Urey HC. The thermodynamic properties of isotopic substances. Journal of the Chemical Society. 1947;**0**:562-581. DOI: 10.1039/jr9470000562

[15] Dansgaard W. Stable isotopes in precipitation. Tellus. 1964;**16**:436-468. DOI: 10.1111/j.2153-3490.1964.tb00181.x

[16] Jouzel J, Merlivat L. Deuterium and oxygen 18 in precipitation: Modeling of the isotopic effects during snow formation. Journal of Geophysical Research. 1984;**89**:11749. DOI: 10.1029/JD089iD07p11749

[17] Muller CL, Baker A, Fairchild IJ, Kidd C, Boomer I. Intra-event trends in stable isotopes: Exploring midlatitude precipitation using a vertically pointing micro rain radar. Journal of Hydrometeorology. 2015;**16**:194-213. DOI: 10.1175/JHM-D-14-0038.1

[18] Yoshimura K. Stable water isotopes in climatology, meteorology, and hydrology: A review. Journal of the Meteorological Society of Japan. 2015;**93**:513-533. DOI: 10.2151/jmsj.2015-036

[19] Craig H, Gordon LI. Deuterium and oxygen-18 variations in the ocean and the marine atmosphere. In: Tongiorgi E, editor. Stable Isotopes in Oceanographic Studies and Paleotemperatures. Spoleto: Conferences in Nuclear Geology; 1965

[20] Rozanski K, Araguás-Araguás L, Gonfiantini R. Isotopic patterns in modern global precipitation. In: Swart PK, Lohmann KC, Mckenzie J, Savin S, editors. Geophysical Monograph Series. Washington, D.C.: American Geophysical Union; 1993. pp. 1-36. DOI: 10.1029/GM078p0001

[21] Celle-Jeanton H, Gonfiantini R, Travi Y, Sol B. Oxygen-18 variations of rainwater during precipitation: Application of the Rayleigh model to selected rainfalls in southern France. Journal of Hydrology. 2004;**289**:165-177. DOI: 10.1016/j.jhydrol.2003.11.017

[22] Gedzelman SD, Lawrence JR. The isotopic composition of Precipitatin from two extratropical cyclones. American Meteorological Society. 1990;**118**:495-509. DOI: 10.1175/1520-0493(1990)118,0495: TICOPF.2.0.CO;2

[23] Graf P, Wernli H, Pfahl S, Sodemann H. A new interpretative framework for below-cloud effects on stable water isotopes in vapour and rain. Atmospheric Chemistry and Physics. 2019;**19**:747-765. DOI: 10.5194/acp-19-747-2019

[24] Lee J-E, Fung I. "Amount effect" of water isotopes and quantitative analysis of post-condensation processes. Hydrological Processes. 2008;**22**:1-8. DOI: 10.1002/hyp.6637

[25] Srivastava R, Ramesh R, Rao TN. Relationship between stable isotope ratios and drop size distribution in tropical rainfall. Journal of Atmospheric Chemistry. 2012;**69**:23-31. DOI: 10.1007/s10874-012-9227-4

[26] Miyake Y, Matsubaya O, Nishihara C. An isotopic study on meteoric precipitation. Papers Meteorology and Geophysics. 1968;**19**:243-266

[27] Stewart MK. Stable isotope fractionation due to evaporation and isotopic exchange of falling waterdrops: Applications to atmospheric processes and evaporation of lakes. Journal of Geophysical Research. 1975;**80**:1133-1146. DOI: 10.1029/JC080i009p01133

[28] Cole JE, Rind D, Webb RS, Jouzel J, Healy R. Climatic controls on interannual variability of precipitation $\delta^{18}O$: Simulated influence of temperature, precipitation amount, and vapor source region. Journal of Geophysical Research. 1999;**104**:14223-14235. DOI: 10.1029/1999JD900182

[29] Risi C, Noone D, Frankenberg C, Worden J. Role of continental recycling in intraseasonal variations of continental moisture as deduced from model simulations and water vapor isotopic measurements: Continental recycling and water isotopes. Water Resources Research. 2013;**49**:4136-4156. DOI: 10.1002/wrcr.20312

[30] Putman AL, Fiorella RP, Bowen GJ, Cai Z. A global perspective on local meteoric water lines: Meta-analytic insight into fundamental controls and practical constraints. Water Resources Research. 2019;**55**:6896-6910. DOI: 10.1029/2019WR025181

[31] Craig H. Isotopic variations in meteoric waters. Science, New Series. 1961;**133**:1702-1703

[32] Marchina C, Zuecco G, Chiogna G, Bianchini G, Carturan L, Comiti F, et al. Alternative methods to determine the δ^2H-$\delta^{18}O$ relationship: An application to different water types. Journal of Hydrology. 2020;**587**:124951. DOI: 10.1016/j.jhydrol.2020.124951

[33] Ehleringer JR, Dawson TE. Water uptake by plants: Perspectives from stable isotope composition. Plant, Cell & Environment. 1992;**15**:1073-1082. DOI: 10.1111/j.1365-3040.1992.tb01657.x

[34] Merlivat L, Jouzel J. Global climatic interpretation of the deuterium-oxygen 18 relationship for precipitation. Journal of Geophysical Research. 1979;**84**:5029. DOI: 10.1029/JC084iC08p05029

[35] Froehlich K, Gibson JJ, Aggarwal P. Deuterium excess in precipitation and its climatological significance. International conference on study of environmental change using isotope techniques; (IAEA-CSP--13/P). Vienna, Austria: International Atomic Energy Agency (IAEA); 2002. Available from: http://www-pub.iaea.org/MTCD/publications/PDF/CSP-13-P_web.pdf/

[36] Risi C, Bony S, Vimeux F. Influence of convective processes on the isotopic composition ($\delta^{18}O$ and δD) of precipitation and water vapor in the tropics: 2. Physical interpretation of the amount effect. Journal of Geophysical Research. 2008;**113**:D19306. DOI: 10.1029/2008JD009943

[37] Santarosa LV, Gastmans D, Sánchez-Murillo R, Santos VD, Batista LV, Betancur SB. Stable isotopes reveal groundwater to river connectivity in a mesoscale subtropical watershed. Isotopes in Environmental and Health Studies. 2021;**57**:236-253. DOI: 10.1080/10256016.2021.1877701

[38] Wassenaar LI, Coplen TB, Aggarwal PK. Approaches for achieving long-term accuracy and precision of $\delta^{18}O$ and δ^2H for waters analyzed using laser absorption spectrometers. Environmental Science & Technology. 2014;**48**:1123-1131. DOI: 10.1021/es403354n

[39] Reboita MS, Gan MA, Rocha RPD, Ambrizzi T. Regimes de precipitação na América do Sul: uma revisão bibliográfica. Revista Brasileira de Meteorologia. 2010;**25**:185-204. DOI: 10.1590/S0102-77862010000200004

[40] Marengo JA, Soares WR, Saulo C, Nicolini M. Climatology of the low-level jet east of the Andes as derived from the NCEP–NCAR reanalyses: Characteristics and temporal variability. Journal of Climate. 2004;**17**:2261-2280. DOI: 10.1175/1520-0442(2004)017<2261:COTLJE>2.0.CO;2

[41] RenéD G. Cold air incursions over subtropical South America: Mean

structure and dynamics. Monthly Weather Review. 2000;**128**:2544-2559. DOI: 10.1175/1520-0493(2000)128 <2544:CAIOSS>2.0.CO;2

[42] Kodama Y. Large-scale common features of subtropical precipitation zones (the Baiu frontal zone, the SPCZ, and the SACZ) part I: Characteristics of subtropical frontal zones. Journal of the Meteorological Society of Japan. 1992;**70**:813-836. DOI: 10.2151/jmsj1965.70.4_813

[43] Stein AF, Draxler RR, Rolph GD, Stunder BJB, Cohen MD, Ngan F. NOAA's HYSPLIT atmospheric transport and dispersion Modeling system. Bulletin of the American Meteorological Society. 2015;**96**:2059-2077. DOI: 10.1175/BAMS-D-14-00110.1

[44] Santos VD, Gastmans D, Santarosa LV, Batista LV, Betancur SB, Dias De Oliverira ME, et al. Variabilidade da Composição Isotópica da Precipitação na Região Central do Estado de São Paulo. Revista Águas Subterrâneas. 2019;**33**:171-181. DOI: 10.14295/ras.v33i2.29474

[45] Dos Santos V, Gastmans D, Sánchez-Murillo R, Felippe Gozzo L, Vianna Batista L, Lilla Manzione R, et al. Regional atmospheric dynamics govern interannual and seasonal stable isotope composition in southeastern Brazil. Journal of Hydrology. 2019;**579**:124136. DOI: 10.1016/j.jhydrol.2019.124136

[46] Gastmans D, Santos V, Galhardi JA, Gromboni JF, Batista LV, Miotlinski K, et al. Controls over spatial and seasonal variations on isotopic composition of the precipitation along the central and eastern portion of Brazil. Isotopes in Environmental and Health Studies. 2017;**53**:518-538. DOI: 10.1080/10256016.2017.1305376

[47] Dos Santos V, Marshall Fleming P, Henrique Mancini L, Dalva Santos Cota S, De Lima GB, Rodrigues Gomes R, et al. Distinguishing the regional atmospheric controls on precipitation isotopic variability in the central-southeast portion of Brazil. Advances in Atmospheric Sciences. 2022;**39**:1693-1708. DOI: 10.1007/s00376-022-1367-0

[48] Santos V, Dias De Oliveira M, Boll J, Sánchez-Murillo R, Menegário AA, Gozzo LF, et al. Isotopic composition of precipitation during strong El Niño–southern oscillation events in the southeast region of Brazil. Hydrological Processes. 2019;**33**:647-660. DOI: 10.1002/hyp.13351

[49] Aggarwal PK, Romatschke U, Araguas-Araguas L, Belachew D, Longstaffe FJ, Berg P, et al. Proportions of convective and stratiform precipitation revealed in water isotope ratios. Nature Geoscience. 2016;**9**:624-629. DOI: 10.1038/ngeo2739

[50] Munksgaard NC, Kurita N, Sánchez-Murillo R, Ahmed N, Araguas L, Balachew DL, et al. Data descriptor: Daily observations of stable isotope ratios of rainfall in the tropics. Scientific Reports. 2019;**9**:14419. DOI: 10.1038/s41598-019-50973-9

[51] Santos V. Isótopos estáveis da água ($\delta^{18}O$-$\delta^{2}H$) em intra-eventos: Decifrando a história da chuva na porção central do estado de São Paulo. [Doctoral Thesis]. Rio Claro: São Paulo State University (UNESP), Institute of Geosciences and Exact Sciences; 2023. Available from: https://repositorio.unesp.br/items/7e021a25-304f-4db5-85ad-c043dd0c7538

Chapter 4

pH Precursors as a Factor for Assessing Rainwater Quality in Roofing Sheets: A Case Study of Rivers State, Nigeria

Daniel O. Omokpariola, John K. Nduka and Patrick L. Omokpariola

Abstract

Rainwater harvesting is vital for water management in water-scarce regions. This study in Rivers State, Nigeria, assesses rainwater quality from different roofing sheets, emphasizing critical pH precursors. Results show a pH range from 4.50 to 7.90, shifting toward alkalinity with increased rainfall. Temperature rises steadily, while conductivity, turbidity, total dissolved solids, and suspended solids decrease during the wet season. Anionic composition showed that Rumuodomaya/Rumuodome had a high level of 4.77 mg/L nitrate, 1.32 mg/L nitrite, and 1.15 mg/L phosphate, while Chokocho has a high level of 11.51 mg/L chloride, 6.48 mg/L sulfate and 3.44 mg/L hydrogen carbonate compared to Ogale and Diobu for zinc roof. Light metal composition indicates concentrations of sodium, calcium, potassium, ammonium, and aluminum. The neutralization factor analysis highlights NH_4^+ importance, with Ca^{2+}, Na^+, and Mg^{2+} play significant roles. Hierarchical and factor analysis showed that influences from industrial emissions, agriculture, biomass burning, road construction, limestone mining, soil resuspension, and metabolic processes impact rainwater's ionic composition. pH emerges as critical, reflecting anthropogenic influences. pH is critical in all aspects of ionic influence from anthropogenic sources that can impact the rainwater quality over a long period. So, rainwater must be treated before consumption or usage for domestic purposes.

Keywords: roofing sheets, cations, anions, pH interaction, rainwater, rivers state, Nigeria

1. Introduction

Access to clean and safe drinking water is fundamental to human health and well-being from an ethical viewpoint [1]. In many parts of the world, including Nigeria, inadequate access to potable water sources poses significant challenges, as a result, alternative water sources, such as rainwater harvesting, have gained considerable attention as a sustainable solution to address water scarcity issues [2].

IntechOpen

Rainwater harvesting involves the process of collecting and storing rainwater for various domestic and non-potable uses as it offers several advantages such as its availability and ease of collection (surface or underground), including reduced strain on existing water resources and the potential for cost savings [3–5]. However, rainwater quality is influenced by multiple factors, which are contamination due to microbial load, and physiochemical, and its suitability for consumption depends on the presence of contaminants [6].

Roofing sheets serve as the primary collection surface for rainwater harvesting systems, and play a crucial role in determining the quality of harvested rainwater via the age of the roofing sheet, type of the roofing materials, deposition time and its corresponding intensity [7] in addition to the composition of roofing materials that can introduce various contaminants, including heavy metals, organic compounds, and microbial agents, into the collected rainwater [8–10]. Among these factors, pH precursors have been recognized as an important indicator of water quality, as they can affect the corrosivity and taste of harvested rainwater [11, 12]. The pH of rainwater can be influenced by several factors, such as atmospheric pollutants, airborne particles, and physiochemical interactions with roofing materials [13, 14]. Roofing sheets made of different materials, such as metal, concrete, or asbestos, can leach substances that alter the pH of rainwater [15–17]. These pH precursors can have potential implications for human health, as they may affect the taste, usability, and suitability of rainwater for various applications, including drinking, cooking, and sanitation [18, 19].

Rivers State, located in southern Nigeria (**Figure 1**), experiences significant rainfall throughout the year, making rainwater harvesting a viable option for meeting water demands. However, several research works such as Omokpariola and Omokpariola, [18], Nduka et al. [20] and Omokpariola et al. [21, 22] conducted studies on the quality of rainwater from different roofing materials in the same region without particular focus on the role of pH precursors. Therefore, this research aims to address this knowledge gap by conducting a comprehensive investigation of the impact of pH precursors on rainwater quality in roofing sheets within the context of Rivers State. By evaluating the pH precursors present in rainwater collected from various roofing materials commonly used in Rivers State, we can gain valuable insights into the quality and safety of harvested rainwater. The investigation aims to shed light on the potential implications of pH precursors for human health and environmental sustainability, ultimately contributing to the development of guidelines and best practices for safe and reliable rainwater harvesting systems in the region. The findings of this study will contribute to enhancing our understanding of the factors affecting rainwater quality and provide valuable information for policymakers, researchers, and individuals involved in rainwater harvesting practices.

2. Materials and methods

2.1 Sample collection and storage

Rainwater samples were collected from different roofing surfaces (zinc, aluminum, asbestos and stone-coated) and ambient (open-air) between the period of April

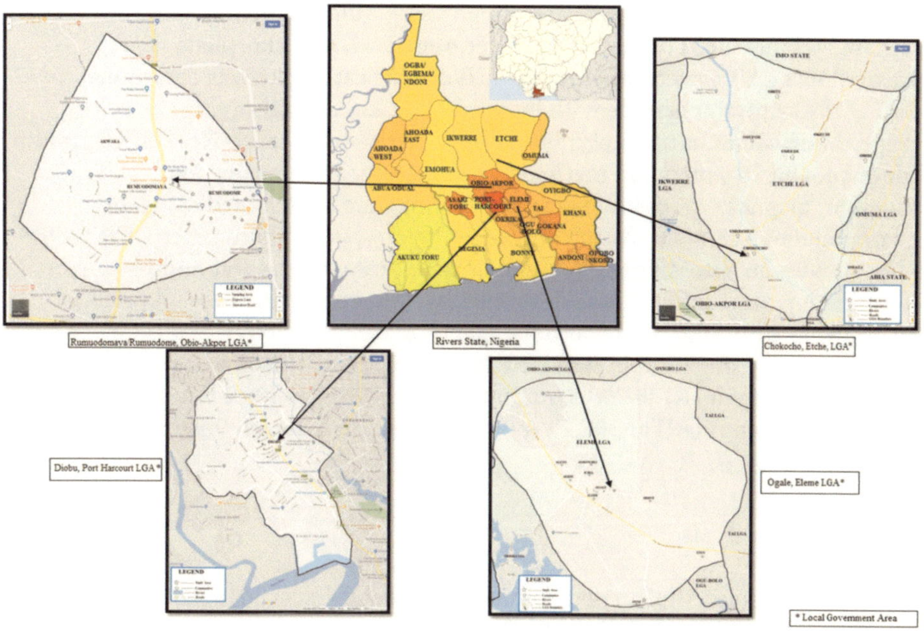

Figure 1.
Sample locations in Rivers state, Nigeria.

to November 2019 as described by Omokpariola et al. [1]. The sampling containers (500 ml) were prewashed, and rinsed severally with rainwater from the sampling location, thereafter, the rainwater samples were placed on support for collection above 1 m using a precleaned bowl across different sampling surfaces respectively. The pH was determined in-situ using a pH meter to give an on-the-spot assessment of rainwater, subsequently, the rainwater was transferred into different sampling containers, labeled accordingly, and packaged in a black cellophane bag for storage and transfer to the laboratory.

2.2 Determination of physiochemical parameters

The methodology used in the study included the determination of the physical and chemical parameters of rainwater samples. The physical parameters measured were pH, temperature, turbidity, conductivity, total dissolved solids, and total suspended solids [23, 24]. For pH determination, a pH meter was calibrated using reference buffer solutions [4, 7, 10], and the pH of rainwater samples was measured in situ. Temperature was determined using a mercury-in-glass thermometer by immersing it in the rainwater sample and recording the reading after equilibrium. Turbidity was measured using a turbidity meter, and conductivity was measured using a conductivity meter that was standardized with a KCl solution. Total dissolved solids (TDS) were determined by filtering a 50 mL rainwater sample, evaporating the filtrate, and weighing the residue. Total suspended solids (TSS) were determined by evaporating the rainwater sample and weighing the total solids. The difference between total solids (TS) and total dissolved solids (TDS) gave the total suspended solids. The chemical parameters analyzed in the rainwater samples included chloride, sulfate, phosphate, hydrogen carbonate, ammonium ion, nitrate, nitrite, and various metallic ions [24, 25].

The chloride concentration was determined using Mohr's argentometric method. Sulphate was measured using the SulfaVer 4 method, and phosphate was measured using UV spectrophotometry. Hydrogen carbonate was determined by titrating the rainwater sample with hydrochloric acid and observing the color change. Ammonium, nitrate, and nitrite were determined using Nessler, Cadmium Reduction, and Diazotization methods, respectively. The metallic ions: sodium, potassium, magnesium, calcium, and aluminum were determined using an Agilent inductive coupled plasma optical emission spectrophotometer (ICP-OES) equipped with an autosampler, simultaneous optical viewing and multi-elemental determination of metallic ions as described by Omokpariola and Omokpariola [18, 26], where pre-digestion was done using concentrated nitric acid, calibration was determined using sample standard at absorbance values of sodium (Na) at 589.6 nm, potassium (K) at 766.5 nm, magnesium (Mg) at 285.2 nm, calcium (Ca) at 315.9 nm, aluminum (Al) at 308.2 nm with regression between 0.9969 and 1.0000 respectively.

2.3 Statistical analysis

Statistical analysis was carried out using Microsoft Excel 2019, XRealStats Excel Add-ins and Past: paleontological statistics software to determine the mean, standard deviation, range, coefficient of variance, and graphical plots (clustering) of the parameters obtained over three sampling periods for the various rainwater sampled were presented [27–30].

2.4 Neutralization factor

The neutralization factor (NF) is used to indicate the degree of neutralization of cations by specific anion species in atmospheric precipitation [30].

$$\mathrm{NF(x)} = \frac{\mathrm{C_x}}{\mathrm{C_{Cl^-}} + \mathrm{C_{SO_4^{2-}}} + \mathrm{C_{NO_3^-}} + \mathrm{C_{NO_2^-}} + \mathrm{C_{PO_4^{3-}}} + \mathrm{C_{HCO_3^-}}} \tag{1}$$

Where C_x is the concentration of cation (x) of interest.

The role of Na^+, K^+, Mg^{2+}, Ca^{2+}, NH_4^+, Al^{3+} plays an important role in neutralization with anion (x) of interest.

2.5 Factor analysis

Factor analysis (Principal component analysis) has been widely used in different studies such as [30]. The principal component analysis is based on the mathematical model used to analyze multidimensional data to correlate variables thereby extracting salient information. Factor analyses were done using the principal component method using Excel, 2019 – XRealStats Add-in Package for Microsoft Windows. It was done to determine the factor underlying the inter-correlation between measured species. According to Kovacs *et al.*, [31], absolute values of factor loading higher than 0.71 are indicated as possible indicators. The physiochemical parameters (pH; Temperature; Conductivity; Turbidity; TDS; TSS; Cl^-, SO_4^{2-}, NO_3^-, NO_2^-, PO_4^{3-}, HCO_3^-, Na^+, K^+, Mg^{2+}, Ca^{2+}, NH_4^+, and Al^{3+}) were taken into considerations in the factor analysis. Initial factors were extracted for sampling surfaces from different study area in Rivers state, Nigeria. Factor with eigenvalue greater than 1 were considered for varimax rotation to obtain the final matrix for possible interpretation of possible sources.

3. Result

The study assessed the physiochemical and metallic concentration in Ogale, Eleme local government area, Rumuodomaya/Rumuodome, Obio-Akpor local government area, Diobu, Port-Harcourt local government area, Chokocho, Etche local government area of Rivers State for the different monthly rain period.

3.1 Physical parameters of rainwater samples

Table 1 shows mean and standard deviation (SD) of physical parameters from rainwater sampling surfaces (ambient, zinc roofing, aluminum roofing, asbestos roofing, and stone-coated roofing) different locations respectively. The values of pH obtained ranged from 4.50 to 7.90 for different sampling surfaces in all locations (Ogale, Rumuodomaya/Rumuodome, Diobu and Chokocho) respectively, which showed sharp increase from acidity to alkalinity. In relation to sampling surfaces, the sum concentration of pH for Ogale and Diobu reveals that asbestos roofing sheet was highest, while zinc roofing was lowest. Rumuodomaya/Rumuodome and Chokocho reveals that Stone-coated roofing sheet was highest compared to aluminum roofing sheet that was lowest. The values of temperature obtained ranged from 18.70 to 25.90° C for different sampling surfaces in all locations. Temperature increased from early rain (April) to late rain (November) regiments as mean temperature for all location was zinc roof gave highest, while ambient was lowest except for Rumuodomaya/

Sampling surface type	Type of parameter	Ogale, Eleme LGA	Rumuodomaya/ Rumuodome, Obio-Akpor LGA	Diobu, Port-Harcourt LGA	Chokocho, Etche LGA
N: Number of Samplings		Mean ± SD	Mean ± SD	Mean ± SD	Mean ± SD
Ambient *N*: 6 *N*: 11 *N*: 7 *N*: 8	Temp. (°C)	23.81 ± 1.44	24.03 ± 1.61	24.04 ± 1.11	24.19 ± 0.96
	pH	6.41 ± 0.91	6.83 ± 0.72	6.72 ± 0.55	6.67 ± 0.69
	Cond. (μS/cm)	106.43 ± 173.75	45.13 ± 33.28	66.85 ± 90.65	126.51 ± 143.64
	Turbidity (NTU)	1.54 ± 1.18	2.89 ± 1.81	1.84 ± 1.18	0.97 ± 0.34
	TDS (mg/L)	10.14 ± 5.76	19.36 ± 11.00	13.61 ± 9.65	12.93 ± 10.39
	TSS (mg/L)	2.88 ± 2.70	2.81 ± 1.71	2.51 ± 1.54	3.66 ± 3.11
	Cl^- (mg/L)	2.58 ± 5.01	3.02 ± 5.12	6.75 ± 7.18	11.79 ± 12.09
	SO_4^{2-} (mg/L)	1.57 ± 1.66	1.97 ± 1.03	3.14 ± 3.60	5.09 ± 4.63
	NO_3^- (mg/L)	3.08 ± 2.62	2.87 ± 2.38	2.25 ± 1.73	3.39 ± 3.12
	NO_2^- (mg/L)	0.98 ± 1.04	0.64 ± 0.39	0.35 ± 0.22	0.72 ± 0.46
	PO_4^{3-} (mg/L)	0.44 ± 0.29	0.82 ± 0.43	0.51 ± 0.39	0.49 ± 0.41
	HCO_3^- (mg/L)	1.89 ± 1.66	1.31 ± 0.50	0.64 ± 0.36	0.63 ± 0.59
	Na^+ (mg/L)	1.09 ± 1.42	0.81 ± 0.96	0.75 ± 1.00	0.97 ± 2.04
	K^+ (mg/L)	0.26 ± 0.31	0.17 ± 0.28	0.15 ± 0.16	0.18 ± 0.37
	Mg^{2+} (mg/L)	0.33 ± 0.30	0.26 ± 0.35	0.07 ± 0.05	0.09 ± 0.04

Sampling surface type	Type of parameter	Ogale, Eleme LGA	Rumuodomaya/ Rumuodome, Obio-Akpor LGA	Diobu, Port-Harcourt LGA	Chokocho, Etche LGA
	Ca^{2+} (mg/L)	0.67 ± 0.60	0.37 ± 0.54	0.18 ± 0.09	0.23 ± 0.10
	Al^{3+} (mg/L)	0.04 ± 0.06	0.07 ± 0.07	0.02 ± 0.03	0.01 ± 0.01
	NH_4^+ (mg/L)	4.83 ± 3.68	5.38 ± 3.94	3.97 ± 2.50	5.55 ± 3.77
Zinc Roof *N*: 2 *N*: 3 *N*: 3 *N*: 2	Temp. (°C)	24.06 ± 1.31	23.90 ± 2.10	24.14 ± 1.14	24.40 ± 0.90
	pH	6.20 ± 1.03	6.70 ± 0.51	6.51 ± 0.53	6.09 ± 1.05
	Cond. (μS/cm)	115.43 ± 216.76	79.09 ± 62.83	144.91 ± 213.04	194.11 ± 244.71
	Turbidity (NTU)	1.60 ± 1.43	3.45 ± 1.78	2.70 ± 1.32	1.67 ± 0.67
	TDS (mg/L)	18.89 ± 19.58	27.64 ± 16.24	20.01 ± 19.03	19.01 ± 17.48
	TSS (mg/L)	6.36 ± 7.73	4.79 ± 4.01	6.07 ± 9.66	7.63 ± 9.98
	Cl^- (mg/L)	3.36 ± 6.29	5.40 ± 10.00	12.27 ± 16.41	13.94 ± 17.08
	SO_4^{2-} (mg/L)	3.06 ± 2.94	3.66 ± 3.00	3.71 ± 4.02	7.61 ± 8.12
	NO_3^- (mg/L)	4.75 ± 4.01	4.76 ± 5.67	1.86 ± 1.21	5.10 ± 5.88
	NO_2^- (mg/L)	1.29 ± 1.10	0.83 ± 0.50	0.55 ± 0.27	0.99 ± 0.62
	PO_4^{3-} (mg/L)	0.81 ± 0.42	1.74 ± 1.74	3.00 ± 4.20	4.04 ± 5.25
	HCO_3^- (mg/L)	2.71 ± 2.00	1.47 ± 1.09	1.93 ± 2.57	2.36 ± 1.96
	Na^+ (mg/L)	1.44 ± 2.05	1.13 ± 1.58	0.89 ± 1.03	2.29 ± 5.08
	K^+ (mg/L)	0.34 ± 0.44	0.56 ± 0.72	0.26 ± 0.32	0.17 ± 0.28
	Mg^{2+} (mg/L)	0.51 ± 0.42	0.33 ± 0.35	0.12 ± 0.04	0.22 ± 0.23
	Ca^{2+} (mg/L)	1.07 ± 0.93	0.51 ± 0.63	0.46 ± 0.23	0.29 ± 0.15
	Al^{3+} (mg/L)	0.08 ± 0.10	0.21 ± 0.32	0.04 ± 0.03	0.04 ± 0.03
	NH_4^+ (mg/L)	6.89 ± 4.86	7.35 ± 6.40	5.79 ± 5.85	5.82 ± 3.31
Aluminum Roof *N*: 2 *N*: 3 *N*: 3 *N*: 2	Temp. (°C)	23.80 ± 1.55	23.91 ± 2.22	24.01 ± 1.04	24.37 ± 0.70
	pH	6.24 ± 0.97	6.49 ± 0.71	6.46 ± 0.63	6.30 ± 0.85
	Cond. (μS/cm)	139.75 ± 255.48	70.02 ± 87.89	123.61 ± 169.53	116.74 ± 122.84
	Turbidity (NTU)	2.02 ± 1.47	3.43 ± 2.87	2.21 ± 1.15	1.80 ± 0.70
	TDS (mg/L)	23.68 ± 29.91	27.08 ± 14.37	12.91 ± 8.20	24.01 ± 33.73
	TSS (mg/L)	7.21 ± 11.48	4.24 ± 3.13	4.54 ± 6.71	8.46 ± 14.69
	Cl^- (mg/L)	5.23 ± 10.65	3.15 ± 4.79	12.29 ± 17.45	9.98 ± 8.93
	SO_4^{2-} (mg/L)	3.55 ± 4.16	2.81 ± 1.33	4.53 ± 4.44	4.90 ± 5.18
	NO_3^- (mg/L)	5.42 ± 5.94	3.65 ± 3.91	1.86 ± 1.35	5.74 ± 7.60
	NO_2^- (mg/L)	1.30 ± 1.29	0.64 ± 0.32	0.54 ± 0.45	1.06 ± 0.91
	PO_4^{3-} (mg/L)	0.75 ± 0.41	1.23 ± 0.41	0.81 ± 0.53	0.45 ± 0.30
	HCO_3^- (mg/L)	2.79 ± 2.07	1.78 ± 1.22	3.25 ± 5.83	4.31 ± 7.74
	Na^+ (mg/L)	1.55 ± 2.16	0.84 ± 1.13	0.95 ± 0.99	2.30 ± 4.98
	K^+ (mg/L)	0.50 ± 0.81	0.37 ± 0.42	0.17 ± 0.08	0.30 ± 0.50
	Mg^{2+} (mg/L)	0.46 ± 0.51	0.40 ± 0.53	0.18 ± 0.11	0.43 ± 0.35
	Ca^{2+} (mg/L)	1.05 ± 0.93	1.11 ± 1.19	0.47 ± 0.34	0.58 ± 0.52
	Al^{3+} (mg/L)	0.13 ± 0.17	0.28 ± 0.35	0.08 ± 0.05	0.07 ± 0.04
	NH_4^+ (mg/L)	7.71 ± 6.97	6.08 ± 4.95	4.95 ± 4.38	6.28 ± 3.01

Sampling surface type	Type of parameter	Ogale, Eleme LGA	Rumuodomaya/ Rumuodome, Obio-Akpor LGA	Diobu, Port-Harcourt LGA	Chokocho, Etche LGA
Asbestos Roof *N*: 2 *N*: 3 *N*: 2 *N*: 2	Temp. (°C)	23.86 ± 1.38	24.10 ± 1.59	24.13 ± 1.12	24.36 ± 0.91
	pH	6.47 ± 1.00	6.80 ± 0.52	6.60 ± 0.58	6.40 ± 1.12
	Cond. (μS/cm)	121.58 ± 215.52	127.51 ± 172.43	134.45 ± 195.06	156.56 ± 175.67
	Turbidity (NTU)	1.52 ± 1.38	3.93 ± 2.00	2.20 ± 1.16	1.70 ± 0.70
	TDS (mg/L)	19.87 ± 22.91	31.31 ± 19.67	14.56 ± 7.09	21.51 ± 21.98
	TSS (mg/L)	4.76 ± 7.02	4.80 ± 4.44	3.71 ± 2.90	9.39 ± 16.55
	Cl^- (mg/L)	3.55 ± 6.81	4.46 ± 7.68	10.32 ± 13.74	12.92 ± 14.92
	SO_4^{2-} (mg/L)	3.01 ± 3.06	3.22 ± 2.33	4.02 ± 3.48	6.62 ± 6.54
	NO_3^- (mg/L)	3.51 ± 4.30	3.64 ± 4.49	2.71 ± 2.01	4.76 ± 6.60
	NO_2^- (mg/L)	1.04 ± 1.23	0.85 ± 0.55	0.57 ± 0.40	0.86 ± 0.49
	PO_4^{3-} (mg/L)	0.85 ± 0.45	1.43 ± 0.40	1.12 ± 0.92	0.56 ± 0.60
	HCO_3^- (mg/L)	2.32 ± 1.84	3.25 ± 5.31	3.71 ± 6.56	4.87 ± 7.33
	Na^+ (mg/L)	1.50 ± 2.24	1.29 ± 2.14	1.09 ± 1.32	2.52 ± 5.44
	K^+ (mg/L)	0.42 ± 0.65	0.60 ± 0.78	0.38 ± 0.36	0.41 ± 0.67
	Mg^{2+} (mg/L)	0.44 ± 0.45	0.48 ± 0.40	0.36 ± 0.43	0.30 ± 0.34
	Ca^{2+} (mg/L)	1.13 ± 1.62	0.94 ± 1.40	0.57 ± 0.49	0.49 ± 0.31
	Al^{3+} (mg/L)	0.19 ± 0.36	0.27 ± 0.38	0.05 ± 0.02	0.06 ± 0.06
	NH_4^+ (mg/L)	6.20 ± 5.80	6.30 ± 4.20	4.85 ± 3.57	4.56 ± 2.52
Stone-Coated Roof *N*: 0 *N*: 2 *N*: 0 *N*: 2	Temp. (°C)		23.91 ± 2.43		24.36 ± 0.98
	pH		6.99 ± 0.25		6.68 ± 0.58
	Cond. (μS/cm)		52.21 ± 27.65		42.86 ± 33.49
	Turbidity (NTU)		3.04 ± 1.44		1.33 ± 0.91
	TDS (mg/L)		21.97 ± 11.13		17.46 ± 19.48
	TSS (mg/L)		4.17 ± 3.39		4.57 ± 5.36
	Cl^- (mg/L)		3.19 ± 5.67		10.30 ± 12.31
	SO_4^{2-} (mg/L)		3.14 ± 1.88		2.63 ± 1.14
	NO_3^- (mg/L)		3.04 ± 2.69		5.53 ± 6.26
	NO_2^- (mg/L)		0.65 ± 0.43		0.93 ± 0.63
	PO_4^{3-} (mg/L)		0.90 ± 0.45		0.34 ± 0.30
	HCO_3^- (mg/L)		1.63 ± 0.86		2.56 ± 2.24
	Na^+ (mg/L)		0.68 ± 1.31		1.24 ± 2.89
	K^+ (mg/L)		0.28 ± 0.39		0.22 ± 0.41
	Mg^{2+} (mg/L)		0.54 ± 1.01		0.24 ± 0.13
	Ca^{2+} (mg/L)		1.16 ± 2.20		0.49 ± 0.50
	Al^{3+} (mg/L)		0.23 ± 0.29		0.04 ± 0.03
	NH_4^+ (mg/L)		4.91 ± 3.07		6.10 ± 3.67

N: *Number of samplings across different sampling locations.*

Table 1.
Physiochemical composition of rainwater samples from various locations and sampling surfaces.

Rumuodome where stone-coated roof was higher. Electrical conductivity values ranged from 8.71 to 714.20 μS/cm for different sampling surfaces. There was a sharp decrease in conductivity as reading as rainfall events takes place from early to late rain across all sampling surfaces. The mean concentration of electrical conductivity for location of study reveals that for Ogale, aluminum roof was highest compared to ambient surface; for Rumuodomaya/Rumuodome assessed, asbestos roof was highest, while ambient was least; for Diobu, zinc roof was highest, while ambient was least and Chokocho, zinc roof was highest compared to stone-coated roof.

The turbidity value ranged from 0.41 to 5.25 NTU, for different sampling surfaces. The mean turbidity for all locations reveals that for Ogale, aluminum roof was highest compared to asbestos roof, which was the least, for Rumuodomaya/Rumuodome, Diobu and Chokocho, asbestos, zinc and aluminum roofs was highest, while ambient was least. The values of TDS obtained ranged from 3.45 to 98.12 mg/L for different sampling surfaces. The mean TDS assessed for all locations revealed that ambient was least, while aluminum roof was higher for Ogale and Chokocho, Rumuodomaya/ Rumuodome and Diobu produced higher TDS values for asbestos and zinc roof. The values of TSS obtained ranged from 0.51 to 40.61 mg/L, as the mean TSS assessed in terms of sampling surfaces revealed that zinc roof was highest in TSS concentration compared to ambient for Ogale, Rumuodomaya/Rumuodome and Diobu, while Chokocho was highest for asbestos roof compared to ambient that was least.

3.2 Anionic composition of rainwater

Table 1 shows the mean and standard deviation (SD) of anionic composition from rainwater samples from different locations respectively. Chloride values ranged from 0.11 to 50.80 mg/L for different sampling surfaces in all locations. The mean chloride concentration revealed that zinc roof was highest for Rumuodomaya/Rumuodome, Diobu and Chokocho, while for Ogale - aluminum roof was highest as ambient was least for all locations assessed. The values of sulfate ranged from 0.40 to 23.97 mg/L for different sampling surfaces and locations respectively. The mean sulfate concentration agreed with chloride assessment for all locations and sampling surfaces. Nitrate values ranged from 0.12 to 22.69 mg/L for different sampling surfaces respectively. The mean nitrate concentration reveals that the aluminum roof was highest for Ogale and Chokocho, zinc and asbestos roof was highest for Rumuodomaya/Rumuodome and Diobu as ambient was least for all locations.

Nitrite values ranged from 0.08 to 5.68 mg/L for different sampling surfaces which was agreed with nitrate assessment for all locations and sampling surfaces. Phosphate values obtained ranged from 0.01 to 2.73 mg/L as the mean concentration of phosphate disclose that asbestos roof was highest, while ambient was least for Ogale, Rumuodomaya/Rumuodome and Diobu, for Chokocho, zinc, and stone-coated roof was highest and least accordingly. Hydrogen carbonate values obtained ranged from 0.15 to 21.83 mg/L, hydrogen carbonate discloses that asbestos roof was highest for Rumuodomaya/Rumuodome, Diobu and Chokocho – aluminum roof was highest for Ogale, while ambient was least for all locations.

3.3 Cationic composition of rainwater

Table 1 shows the mean and standard deviation (SD) of light metals composition from rainwater samples from different locations respectively. The sodium values obtained ranged from 0.004 to 13.76 mg/L for different sampling surfaces respectively. The mean concentration of sodium reveals that asbestos roof was highest for Rumuodomaya/Rumuodome, Diobu and Chokocho, aluminum roof was highest for Ogale, ambient was least for all locations. The potassium values obtained ranged from 0.002 to 2.30 mg/L for different sampling surfaces respectively. The mean potassium concentration agreed with the sodium assessment for all locations and sampling surfaces. The magnesium values obtained ranged from 0.002 to 2.82 mg/L for different sampling surfaces, while the mean concentration of magnesium reveals that zinc, stone-coated, asbestos, and aluminum roofs were highest for Ogale, Rumuodomaya/Rumuodome, Diobu and Chokocho as ambient was least.

The calcium values obtained ranged from 0.001 to 4.76 mg/L for different sampling surfaces in Ogale, Rumuodomaya/Rumuodome, Diobu, and Chokocho respectively, as the mean calcium concentration showed asbestos roof was highest for Ogale and Diobu; stone-coated and aluminum roof was highest for Rumuodomaya/Rumuodome and Chokocho as ambient was least. Ammonium values obtained ranged from 0.009 to 20.22 mg/L, as the aluminum roof was highest for Ogale and Chokocho, zinc roof was highest for Rumuodomaya/Rumuodome and Diobu, while ambient was least for all locations. Aluminum values obtained ranged from 0.0005 to 1.06 mg/L, this showed that asbestos roof was highest for Ogale and Chokocho, as aluminum roof was highest for Rumuodomaya/Rumuodome and Diobu as ambient was least across all locations.

3.4 Neutralization factor

The neutralization factor refers to the ability of certain ions (anions and cations) to influence the pH of a solution when they undergo chemical reactions that affect the concentration of hydrogen ions (H+) or hydroxide ions (OH-) in the solution. **Figure 2** shows the impacts of selected metals to neutralize anions from rainwater samples across different locations determined from Eq. (1). Across the different sampling surfaces, NF shows that NH_4^+ played major role due to release by fertilizers

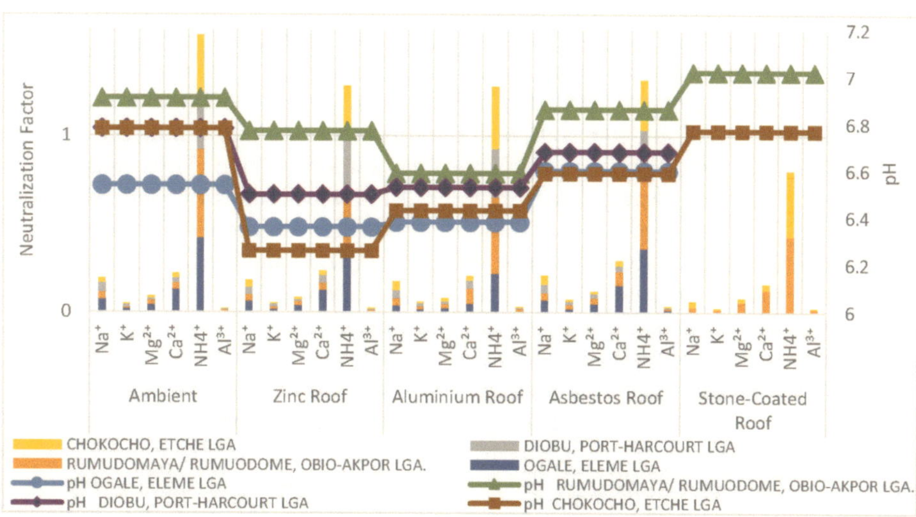

Figure 2.
Neutralization factor of cationic to anionic interactions.

industry and other pollutants from industrial emissions like aerosols or secondary pollutants like sanitary landfills, while Ca^{2+}, Na^{+} and Mg^{2+} contribute to NF by releases from limestone mines and natural soil. K^{+} and Al^{3+} played minor roles in ionic species. This draws to conclude that high concentration of NH_4^{+} with contribution from Ca^{2+}, Na^{+} and Mg^{2+} on neutralization factor, thus raising the pH values in rainwater.

3.5 Hierarchical cluster analysis

Hierarchical clustering involves the process of merging similar data points or clusters into larger clusters until all data points are grouped into a single cluster (agglomerative) or until each data point is its own cluster (divisive), as the data can be visualized as a dendrogram (tree-like structure) [32, 33]. **Figure 3** shows the hierarchical clustering analysis using the correlation coefficient for the set of physiochemical data available for different rainwater surfaces and locations. As shown in **Figure 3a-e**, the data set showed a heat map where conductivity was highest across all rainwater surfaces (ambient, zinc sheets, aluminum sheets, asbestos sheets, stone-coated sheets) respectively, while locations matrix depict that ambient and aluminum roofing sheet data formed similar forming clusters for Ogale, Eleme LGA and Chokocho, Etche LGA and vice versa for zinc and asbestos sheet for Chokocho, Etche LGA and Diobu, Port Harcourt LGA showing similar correlation with Rumuodomaya/Rumuodome Obio-Akpor LGA. The dataset for all rainwater surfaces showed that shorter branches represent a higher similarity between clusters, while longer branches represent a lower similarity, while the heights of the branches indicate levels of dissimilarity or no form of relationship with each merged cluster. Each merged cluster shows a strong or close relationship among themselves.

3.6 Factor analysis

Factor analysis is a statistical technique used to identify underlying factors or dimensions that explain the patterns of correlation among a set of variables, as conducting factor analysis with a rotated varimax rotation, we aim to simplify the data by identifying latent factors that explain the interrelationships among these physiochemical parameters. **Table 2** shows the factor analysis conducted for different roofing sheets in the aforementioned sampled locations shown below with significance levels of $p < 0.05$ and $p < 0.01$ with the threshold for determining the statistical significance of the factor loadings, as **Table 2** gives a total of five (5) varimax factors (F1-F5) with cumulative variance at 73.77%, as F1 is associated with elemental ions and conductivity that suggest the dissolution of minerals and salts from soil and rocks leading to increased total solids (as dissolved solids (TDS) and suspended solids) as shown that contribute to higher conductivity and presence of ions in rainwater. F2 suggests turbiditic influences that are from industrial releases that contribute to particulates in that atmosphere and eventually into rainwater. F3 showed a negative correlation matrix with Ca^{2+}, Mg^{2+}, and Al^{3+} that are due to different polluting or interacting sources from mining operations or marine sprays that affect acidity in rainwater [34]. F4 shows the influence of nitrogen oxides (NOx) from industrial and fossil fuel combustion resulting in acid rain in tandem with ammonia from organic decomposition from waste dumps released into the atmosphere turning to ammonium (NH^{4+}) in the presence of rainwater molecules [22]. F5 represents the combination of pH and temperature that are positively correlated which are due to the potential

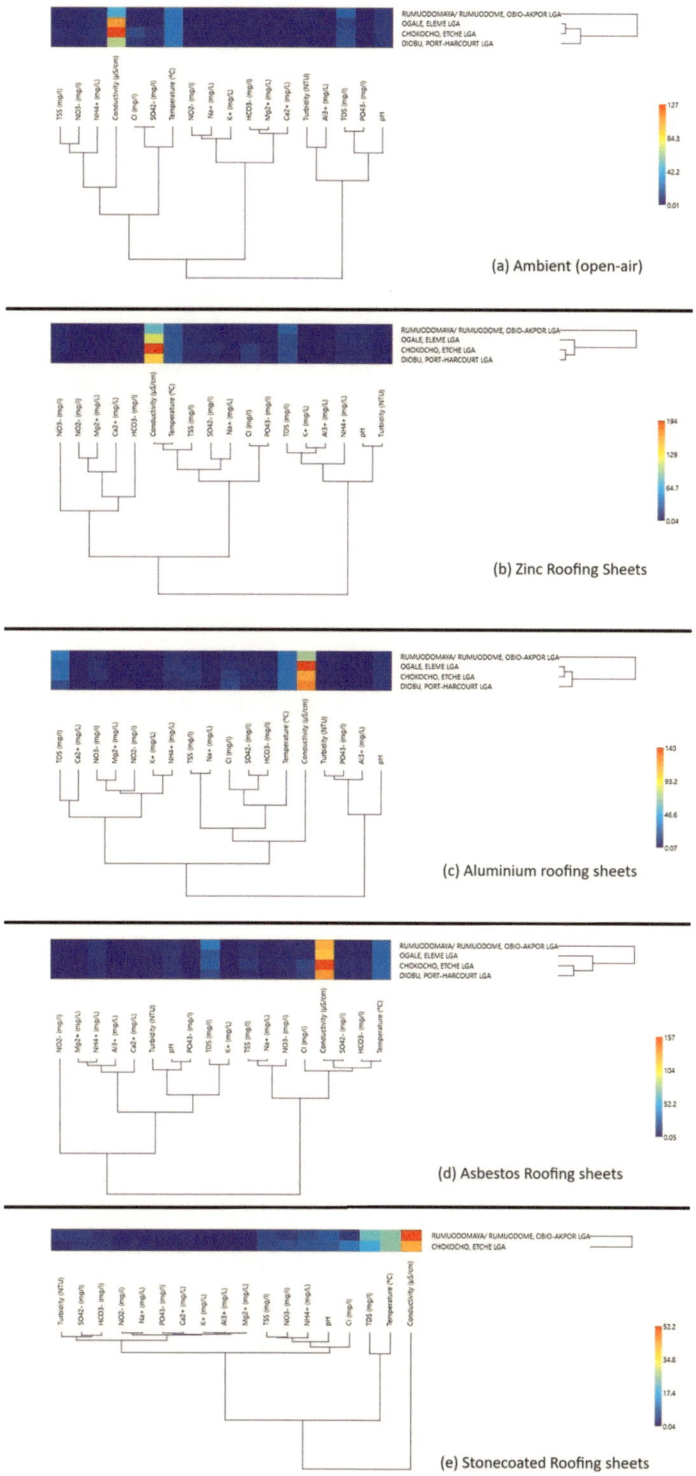

Figure 3.
Hierarchical clustering of rainwater from different sampling surfaces and locations: (a) ambient rainwater, (b) Zinc roofing sheets (c) Aluminium roofing sheet (d) Asbestos roofing sheet, (e) Stone coated roofing sheet.

	F1	F2	F3	F4	F5
Temp. (°C)	−0.27	−0.60	−0.09	−0.58	0.26
pH	−0.32	−0.47	−0.09	−0.64	0.39
Cond.	0.63	0.08	0.16	0.11	−0.49
Turb.	−0.01	0.80	0.07	0.05	−0.47
TDS	0.56	0.67	0.06	0.19	−0.22
TSS	0.85	0.08	0.17	0.06	−0.30
Cl^-	0.94	0.02	0.22	0.03	0.02
SO_4^{2-}	0.92	0.08	0.19	0.00	0.02
NO_3^-	0.68	0.25	0.04	0.58	−0.22
NO_2^-	−0.14	−0.01	−0.06	0.92	−0.01
PO_4^{3-}	−0.05	0.92	0.02	0.09	0.13
HCO_3^-	0.13	0.12	0.00	0.40	−0.82
Na^+	0.78	−0.14	−0.01	−0.04	−0.49
K^+	0.49	0.26	−0.05	−0.07	−0.69
Mg^{2+}	−0.10	−0.15	−0.93	−0.20	0.04
Ca^{2+}	−0.10	−0.11	−0.87	0.14	0.13
Al^{3+}	−0.28	0.14	−0.70	0.01	−0.15
NH_4^+	0.67	0.29	0.13	0.57	−0.15
Eigenvalue	5.23	2.83	2.29	2.54	2.34
Var. (%)	21.32	15.02	12.49	14.12	10.82
Cum (%)	21.32	36.34	48.83	62.95	73.77
Significance, $P \leq 0.05$ *and 0.01* -1 0 1					

Table 2.
Factor analysis of ambient rainwater.

impact of seasonal variation affecting rainwater composition. **Table 3** gives the factor analysis (rotated varimax) of zinc roofing sheets in which the cumulative variance was 61.36% with four (4) factors components, as F1 gave the most variance at 25% with high positive loadings for variance like Conductivity, Total Suspended Solids), Cl^- (Chloride), SO_4^{2-} (Sulfate), NO_3^- (Nitrate), PO_4^{3-} (Phosphate), HCO_3^- (Bicarbonate), Na^+ (Sodium), K^+ (Potassium), and NH_4^+ (Ammonium), which are due to atmospheric deposition of pollutants, urban runoff, and weathering that impacts water quality and ionic conditions, potentially indicating pollution or specific processes affecting zinc roofing sheets. F2 accounts for 15% with positive loading with turbidity and aluminum (Al^{3+}) and negative loading for temperature. F3 and F4 show 10.41% and 10.96% respectively. **Table 4** shows the rotated varimax with 62.43% cumulative variance for aluminum roofing sheets, as F1 gives 21.62% with contributions from both dissolved and suspended solids, ions like Cl^-, SO_4^{2-}, NO_3^-, HCO_3^- and several cations (Na^+, K^+, Mg^{2+}, and NH_4^+) that shows strong correction with conductivity at 0.74% with probable source from industrial discharges, agricultural runoff, and atmospheric deposition of pollutants. F2 produced a variance of 13.93% with strong correction with Mg^{2+}, Ca^{2+}, and Al^{3+} which are associated with

	F1	F2	F3	F4
Temp. (°C)	−0.19	−0.83	−0.16	−0.34
pH	−0.56	−0.20	−0.18	−0.63
Cond. (μS/cm)	0.87	0.23	0.18	0.03
Turb. (NTU)	0.27	0.82	0.25	−0.16
TDS (mg/L)	0.65	0.59	0.06	0.16
TSS (mg/L)	0.95	0.06	0.08	0.05
Cl^- (mg/L)	0.94	−0.01	0.19	−0.02
SO_4^{2-} (mg/L)	0.92	0.04	0.14	0.18
NO_3^- (mg/L)	0.74	0.19	−0.05	0.41
NO_2^- (mg/L)	0.06	0.01	0.06	0.93
PO_4^{3-} (mg/L)	0.89	−0.04	0.12	0.02
HCO_3^- (mg/L)	0.77	0.19	0.09	0.38
Na^+ (mg/L)	0.88	0.00	−0.10	0.14
K^+ (mg/L)	0.65	0.47	−0.22	0.00
Mg^{2+} (mg/L)	0.00	−0.02	−0.93	0.02
Ca^{2+} (mg/L)	−0.18	−0.11	−0.82	−0.19
Al^{3+} (mg/L)	−0.22	0.81	−0.14	0.04
NH_4^+ (mg/L)	0.67	0.40	0.07	0.36
Eigenvalue	7.91	2.93	1.87	1.97
Variance (%)	25.00	15.00	10.41	10.96
Cumulative (%)	25.00	40.00	50.40	61.36
Significance, P ≤ *0.05 and 0.01* -1 0 1				

Table 3.
Factor analysis of rainwater from zinc sheet.

	F1	F2	F3	F4
Temp. (°C)	−0.16	−0.42	−0.65	−0.50
pH	−0.41	−0.24	−0.50	−0.61
Cond. (μS/cm)	0.74	−0.05	0.55	0.06
Turb. (NTU)	0.26	0.42	0.80	−0.04
TDS (mg/L)	0.88	0.24	0.28	0.10
TSS (mg/L)	0.97	−0.11	0.09	0.04
Cl^- (mg/L)	0.65	−0.41	0.33	0.11
SO_4^{2-} (mg/L)	0.88	−0.17	0.25	0.21
NO_3^- (mg/L)	0.88	0.16	0.02	0.30
NO_2^- (mg/L)	0.07	0.14	−0.16	0.90
PO_4^{3-} (mg/L)	0.00	0.01	0.85	−0.06
HCO_3^- (mg/L)	0.84	−0.10	−0.02	0.22

	F1	F2	F3	F4
Na^+ (mg/L)	0.96	0.08	−0.14	0.07
K^+ (mg/L)	0.85	0.07	0.12	−0.10
Mg^{2+} (mg/L)	0.13	0.82	0.10	0.37
Ca^{2+} (mg/L)	−0.13	0.81	0.06	0.02
Al^{3+} (mg/L)	0.00	0.72	0.48	0.10
NH_4^+ (mg/L)	0.60	0.17	0.41	0.57
Eigenvalue	7.23	2.62	3.07	2.13
Variance (%)	21.62	13.93	15.35	11.53
Cumulative (%)	21.62	35.55	50.90	62.43
Significance, $P \leq 0.05$ *and 0.01* -1 0 1				

Table 4.
Factor analysis of rainwater from aluminum sheet.

construction, mining, and soil resuspension [1] that can lead to increasing turbidity and dissolved solids. F3 shows influence from particulates from natural and anthropogenic releases. F4 was influenced by nitrite-related pollutants. **Table 5** shows the factor analysis with the influence of rainwater on asbestos roofing sheets with a cumulative variance of 61.30%. F1 shows 24.47% for a strong correlation with cations and anions from industrial discharges, agricultural runoff, and atmospheric deposition of pollutants affecting asbestos roofing sheets. F2 accounts for 12.63% with a strong relationship for TDS, turbidity, phosphate, and potassium which relate to the presence of specific ions and pollutants, possibly from industrial sources or weathering of roofing materials. F3 and F4 showed a negative correlation due to different

	F1	F2	F3	F4
Temp. (°C)	−0.19	−0.58	0.12	0.64
pH	−0.50	−0.25	−0.09	0.69
Cond. (μS/cm)	0.80	0.45	0.03	−0.06
Turb. (NTU)	0.33	0.78	−0.08	−0.21
TDS (mg/L)	0.76	0.35	0.00	−0.38
TSS (mg/L)	0.94	−0.08	0.03	−0.08
Cl^- (mg/L)	0.82	0.21	0.11	0.03
SO_4^{2-} (mg/L)	0.94	0.04	0.09	−0.16
NO_3^- (mg/L)	0.93	0.11	0.00	−0.19
NO_2^- (mg/L)	0.15	−0.14	0.04	−0.91
PO_4^{3-} (mg/L)	−0.04	0.88	0.13	−0.07
HCO_3^- (mg/L)	0.92	0.01	−0.13	−0.06
Na^+ (mg/L)	0.93	−0.09	−0.05	−0.12
K^+ (mg/L)	0.83	0.32	−0.17	−0.10

DOI: http://dx.doi.org/10.5772/intechopen.1004027

	F1	F2	F3	F4
Mg^{2+} (mg/L)	0.14	0.02	−0.90	−0.21
Ca^{2+} (mg/L)	−0.10	−0.04	−0.88	0.06
Al^{3+} (mg/L)	−0.21	0.29	−0.23	−0.71
NH_4^+ (mg/L)	0.51	0.41	−0.13	−0.62
Eigenvalue	7.68	2.55	1.75	2.94
Variance (%)	24.47	12.63	8.87	15.32
Cumulative (%)	24.47	37.11	45.98	61.30
Significance, P ≤ *0.05 and 0.01* -1 0 1.				

Table 5.
Factor analysis of rainwater from Asbestos sheet.

	F1	F2	F3	F4
Temp. (°C)	−0.14	−0.84	−0.33	0.09
pH	−0.51	−0.12	−0.56	0.00
Cond. (μS/cm)	0.60	0.56	0.38	0.30
Turb. (NTU)	0.15	0.87	0.08	0.04
TDS (mg/L)	0.82	0.46	0.11	0.07
TSS (mg/L)	0.92	−0.04	0.21	0.16
Cl^- (mg/L)	0.96	−0.13	0.11	0.03
SO_4^{2-} (mg/L)	0.04	0.62	0.29	−0.59
NO_3^- (mg/L)	0.93	−0.01	0.04	−0.32
NO_2^- (mg/L)	0.04	−0.16	0.23	−0.86
PO_4^{3-} (mg/L)	0.07	0.86	−0.19	0.10
HCO_3^- (mg/L)	0.95	0.15	0.09	−0.06
Na^+ (mg/L)	0.96	0.01	−0.11	−0.10
K^+ (mg/L)	0.86	0.23	−0.10	−0.16
Mg^{2+} (mg/L)	−0.05	−0.03	−0.92	0.23
Ca^{2+} (mg/L)	0.07	0.01	−0.94	0.22
Al^{3+} (mg/L)	−0.25	0.78	−0.48	0.06
NH_4^+ (mg/L)	0.68	0.01	0.25	−0.65
Eigenvalue	7.06	3.87	2.88	1.90
Variance (%)	25.08	19.36	15.87	10.47
Cumulative (%)	25.08	44.44	60.30	70.77
Significance, P ≤ *0.05 and 0.01.* -1 0 1				

Table 6.
Factor analysis of rainwater from stone-coated sheet.

biotransformation and migration rate. **Table 6** shows the factor analysis of rainwater from stone-coated roofing sheets as cumulative variance was at 70.77%. F1 shows significant loading for variance like TDS (Total Dissolved Solids), TSS (Total

Suspended Solids), Cl^- (Chloride), NO_3^- (Nitrate), HCO_3^- (Bicarbonate), Na^+ (Sodium), K^+ (Potassium), and NH_4^+ (Ammonium) at 25.08% suggesting influences from various anionic and cationic influences leading to increased total solids (dissolved and suspended) and conductivity at 0.60 leading to salt formation and taste change. F2 shows the influence of weathering, agricultural and industrial activities from phosphate (PO_4^{3-}), sulfate (SO_4^{2-}) and aluminum (Al^{3+}). F3 and F4 account for 19.36% and 10.47% of the variance with a negative correlation.

4. Discussion

Having assessed rainwater from varying locations and sampling surfaces, pH steadily increased from acidity to alkalinity by dilution potential over more rainfall events throughout the sampling months, pH is usually a quick indication of water quality as it shows the impact of external contaminants in rainwater. The presence of dissolvable anions such as oxides of sulfur, nitrogen, chloride and phosphates and hydrogen carbonate impacts on pH of rainwater, which agrees with Okpoebo et al. [35]; Gav et al. [36]; Al-Amoush et al. [37] physiochemical assessment, which they attributed to releases from industrial, automobile, and gas flares, which leads to degradation of metallic surfaces over long period. Low pH values influence high reactivity, and dissolution of metallic surfaces, as metallic ions leach out into rainwater influencing taste and color [1]. This constant degradation of metallic surfaces is due to different chemical reactions such as redox, precipitation, hydrolysis, and neutralization, therefore producing metallic ions that are released into rainwater. Although other external factors such as transboundary pollutants, marine contribution, burning of sanitary fields, mining activities and agricultural activities can further initiate reaction and leaching of metallic surfaces thus influencing the esthetic quality of rainwater.

Temperature is an esthetic measure of drinking water, although it conforms to the WHO recommended range of 20–30°C thereby posing no risk to human health [38]. Temperature is due to convectional cooling and heating of the earth's surface, absorbed by different surfaces and dissipated as infrared from anthropogenic inputs such as industrial cooling towers, power plants and gas flares around varying sampling locations in Rivers states. The relatively low temperature is attributed to the time of sampling, which was done in the morning and evening after rainfall events. Elevated temperatures lead to an increase in toxicity and bioaccumulation in organisms [39]. Temperature has a strong influence on the solubility and conductivity of ions, dissolved oxygen and contents and biological activities in rainwater, consequently leading to mineralization (chemical reactions) over a period [40]. Mineralization is usually from cationic and anionic interactions, which form precipitate or dissolved particles. This mineralization of particles forms a rationale for turbidity and total solids (a combination of dissolved solids and suspended solids), which increases ionic mobility, temperature, and pathogen growth from anthropogenic inputs [38–40]. Turbidity and total solids impact the taste, odor and color of rainwater, as consumption of contaminated rainwater has the potential to cause diseases from microorganisms such as diarrhea, dysentery, stomach upset, and purging in humans [38, 41]. The cationic and anionic composition of rainwater influences pH, temperature, conductivity, turbidity, total solids in rainwater quality over a period due to chemical reactions taking place as rainfall deposition takes place with inputs from collection surfaces for domestic and other uses. Chloride ions combine with varying cations such as calcium, magnesium, sodium, potassium, aluminum, and ammonium ions forming salt-based rainwater that impacts the taste of rainwater, which can be attributed to marine and sea-sprays contribution, soil resuspension as seen in neutralization factor (NF) assessment presented in **Figure 3**, One can iterate that pH was influenced from releases of these contaminants in rainwater [42]. Across the different sampling surfaces, NF shows that NH_4^+ played major role due to release by fertilizers industry and other pollutants from industrial emissions like aerosols or secondary pollutants like sanitary landfills, while Ca^{2+}, Na^+ and Mg^{2+} contribute to NF by releases from limestone mines and natural soil. K^+ and Al^{3+} played minor roles to ionic species. This draws to conclude that high concentration of NH_4^+ with contribution from Ca^{2+}, Na^+ and Mg^{2+} on neutralization factor, thus raising the pH values in rainwater [20].

Sulphate, nitrate, nitrite, and phosphates in rainwater gives the eutrophication potential of water quality as it led to microbial growth and releases of obnoxious odor over a period, which was evident in the factor analysis [35]. The oxides of these anions (SOx, NOx, COx, POx) cause climate change, photochemical smog and acid rain from industrial emissions and gas flaring, waste combustions and automobile emission that causes eye irritation, skin reaction with burning sensation, also high concentration of sulfate has laxative and dehydration effect with presence of cations such as magnesium or calcium [42]. Nitrate and nitrites cause methemoglobinemia in bottle-fed babies [38, 41]. Marine contribution is probable with sodium, which therefore shows that sulfate, chloride, nitrate, and nitrite has negative impacts on humans from consumption of rainwater without treatment [43]. Hydrogen carbonate produced correlation with chloride, sulfate and nitrate is attributed to metrological and anthropogenic influences, which increases sediment, turbidity, and taste of rainwater [44].

The wet or dry deposition of anions influences metallic leaching on metallic surfaces over a long period, which thus affects rainwater quality. Sodium, potassium, magnesium, calcium, aluminum, and ammonium plays an important role as it influences, pH, conductivity, turbidity, and total solid potential of water with the influence of anions. They also influence taste, water hardness, dissolved oxygen, and color of water [45, 46]. Magnesium and calcium are an indication of water hardness due to the presence of carbonate, sulfate, and chloride as it was evident in factor analysis conducted, which showed high correlation ranging between 0.63 and 0.99 for all roofing sheet, attributed to influence of soil or dust resuspension from weathering, road construction limestone mining [30]. The correlation of sodium and potassium reveal strong correlation with chlorine, sulfate, nitrate, nitrite, and hydrogen carbonate as seen in all F1 (**Table 2**) of all rainwater surfaces suggesting marine contribution from the Atlantic Ocean, mining activities, soil resuspension, burning activities of biomass, agricultural residues, utilization of charcoal for cooking fuels and industrial emission from gas flares, power plants and heat exchangers that impact pH of rainwater [9, 10, 26].

Ammonium showed strong correlation to sulfate, nitrate, and nitrite, which suggest that in rainwater ammonium sulfate, ammonium nitrate and ammonium nitrite that is due to dissolution of aerosols and secondary pollutant, agriculture, and fertilizer emissions [47]. Ammonia is produced from microbial degradation of organic components in water and atmospheric releases from sanitary landfills, sewage, and industrial effluents, which impacts taste and odor, at lower pH, it combines with rainwater forming ammonium and hydroxide ions [38]. Nitrogen composition in the atmosphere is 78%, which is evident in the high correlation with other cations and anions in the environment. Nitrogen content is divided into organic (Kjeldahl nitrogen, ammonium) and inorganic (nitrate, nitrite, and ammonia) forms that is released by natural and anthropogenic sources, which impacts on the composition of nitrogen. Inorganic nitrogen (NOx) across different sampling surfaces were dominant to interact than organic nitrogen (NH_4^+), due to loss of ammonia gas from sample bottles thereby having lower total nitrogen content. Presence of submicron particles of minerals could have increased pH during transport of samples which led to loss of ammonia (NH_3); although, other anomalies from underestimation from spectroscopic method could have led to lower value of total nitrogen content [43]. Inorganic form of nitrogen is via biomass burning, fossil fuel combustion, soil releases, lightening, aviation, volcanic eruptions etc. Organic forms of nitrogen are from microbial actions of dead matter and industrial effluent waste, which forms ammonia as by-product. Nitrogen oxides combines with water vapor in the atmosphere to form acids which when precipitation takes place, it led to lower pH as seen from the **Table 1** with high concentration of NO_3^-, NO_2^- and NH_4^+ respectively. Due to presence of aerosols from interactions with water vapor and other particulates, there is potential for nitrogen groups to combine forming $(NH_4)HSO_4$, $(NH_4)_2SO_4$, NH_4NO_3 and NH_4NO_2 that can dissolve into cloud droplets forming smog with sulfate groups which has adverse effects on health issues such as asthma attack, difficulty to breath well [48]. According to Ohara, [49], presence of free radicals or reactive ions of anionic oxide can react with ozone causing decrease in ozone content by influence of ultraviolet radiation from the sun.

4.1 Impact of sampling surface on rainwater quality and leaching matrix

Rainwater collected from diverse sampling surfaces (zinc roof, aluminum roof, asbestos roof, and stone-coated roof), it used to calculate its influence on ambient rainwater samples assessed using the formula:

$$\text{Influence of ambient to sampling surfaces } (A - SS) = \frac{C_{x\text{ambient}}}{C_{x\text{roof of interest}}} \times 100\% \quad (2)$$

$$\text{Infuence of external contaminant} = 100 - (A - SS) \quad (3)$$

Where C_x = Concentration of parameters of interest

The concentration of anions and cations with microbial input has great influence on physical parameters of rainwater quality in comparison to different sampling surfaces and location assessed. The presence of particles in the atmosphere released from anthropogenic sources reacts in the atmosphere forming chemical products that are transboundary before wet deposition (rainfall) or dry deposition (soil resuspension and particulate matter) takes place [50, 51]. Several factors impact the physicochemical concentration across sampling surfaces for rainwater samples such as: the number of particles deposited on different surfaces, age of metallic surfaces, rate of deterioration (leaching), rainfall intensity and duration of wet deposition [52]. In addition, one cannot conclude that metal leaching is from respectively surfaces (roofing sheets) as external contaminant from anthropogenic releases has immersed impact on rainwater quality [53, 54].

A preliminary assessment showed that as anions concentration increased, it leads to high degradation (metal leaching) of the diverse sampling surfaces due to its chemical composition as given in **Table 1**, which was prevalent in quantitative assessment of cations in the study. For zinc roof, dissolved zinc metal in relation to ambient rainwater ranged from 41.85 to 84.95% for all locations, which is lower than Heijrick *et al.* [55] who reported between 96 and 99.9%; other metals cumulatively ranged between 0 and 96.67% for all locations. For aluminum roof, dissolved aluminum metal in relation to ambient rainwater samples ranged between 29.10 and 36.19% for all locations, which was lower than chemical composition of aluminum (**Table** 7) between 99.0 and 99.60%, as cumulative metals ranged between 0 and 93.32% respectively. For asbestos roofs, dissolved magnesium metal ranged from 21.99 to 71.12%, which was lower than chemical composition of asbestos that contains about 99.95% of hydrous magnesium silicate, other metals ranged from 0 to 100%. For

Roof Type	Pictorial	Chemical composition	Source
Zinc Roof		99.995% (pure zinc) and 0.005% (trace amount of iron, aluminum, copper and titanium).	[56]
Aluminum Roof		99.00–99.60% (unalloyed aluminum) and ≤ 1% (trace amount of iron, copper, manganese, silicon, magnesium, zinc, and other metallic elements).	[57]
Asbestos Roof		99.95% (hydrous magnesium silicate) and ≤ 0.05% (trace amount of calcium, iron, sodium, and silicate).	[58]
Stone-coated Roof		55% aluminum, 43% zinc coated steel coated with natural/ceramic coated volcanic stone chips and ≤ 2% (trace amount of calcium, iron, and other metallic elements).	[59]

Table 7.
Summary of chemical composition of roofing materials.

	Correlation to Ambient rainwater				Influence of External contaminants			
Sampling surface	**OG**	**R/R**	**DB**	**CK**	**OG**	**R/R**	**DB**	**CK**
Zinc	68.58	64.48	63.39	71.78	31.42	35.52	36.62	28.23
Aluminum	59.63	76.24	64.58	76.13	40.37	24.19	35.42	23.87
Asbestos	74.62	61.5	60.64	75.35	25.38	38.47	39.36	24.65
Stone-coated	—	78.93	—	77.36	—	21.07	—	25.81

OG: Ogale, R/R: Rumuodomaya/Rumuodome, DB: Diobu, CK: Chokocho.

Table 8.
Average of physicochemical parameters.

stone-coated roof, dissolved aluminum for Rumuodomaya/Rumuodome and Chokocho are 33.95% and 36.59%, which is lower than 55% aluminum as stated in **Table 1**, dissolved zinc metal is 87.57% and 93.02% that is higher than 43% zinc composition of stone-coated roof, other metals ranged between 1.44 and 93.02% respectively [60].

The cumulative impact of different sampling surfaces (zinc, aluminum, asbestos and stone-coated) with ambient (open-air) were conducted for various parameters are shown in **Table 8**. Several external issues could influence the results from instrumental error, contaminants deposited before rainfall events, amount of anthropogenic contaminant releases in terms of transboundary movement and atmospheric physiochemical reaction [61–63]. Having assessed these, aluminum and zinc roofs did not leach higher metal concentrations compared to asbestos and stone-coated roofs, and so rainwater must be filtered and treated prior to domestic use. For asbestos, has been categorized by the United States Environmental Protection Agency, USEPA [25] and World Health Organization, WHO [64] as being carcinogenic to human health: as such, rainwater collection from it should be discontinued to prevent possible illness and death for both adults and children.

5. Conclusion

The study analyzed the impact of pH precursors in rainwater quality, which are predominantly from roofing surfaces and physiochemical influence in rainwater matrix that interacts across a range of acidity and alkalinity as seen with the pH range of 4.70–7.90. It is critical to also note that as pH decreases (greater acidity), metal solubility increases and the metal concentrations in runoff also increase and vice-versa that was predominant from early rain period to late rain that is alkaline rainwater, as the physicochemical parameters in compliance to study aims provide the following findings:

1. Most of the physicochemical parameters analyzed from diverse sampling surfaces (ambient, zinc roof, aluminum roof, asbestos roof, and stone-coated roof) showed varying concentrations of rainwater quality.

2. The rainwater quality of the sampling surfaces decreased from early rain to late rain due to more rainfall dilution taking place except for temperature (18.70–25.90°C) that increased in progressive medium, as rainwater collected contained

contaminants, which implies rainwater must be treated before portable and convenient use.

3. The chemical composition and age of roofing materials have a tremendous impact on rainwater quality as anions (Cl^-: 0.12–50.80 mg/L; SO_4^{2-}: 0.40–23.97 mg/L; NO_3^-: 0.08–5.68 mg/L) were deposited in increasing progression or vice versa via wet or dry deposition, it initiates cation (Na^+: 0.004–13.76 mg/L; Mg^{2+}: 0.002–2.82 mg/L; Ca^{2+}: 0.001–4.76 mg/L) leaching on the roofing sheets as it ages.

4. In terms of study locations, as particulate contaminants are deposited on different sampling surfaces from natural and/or anthropogenic releases via geographical and transboundary movement, it increases total solids (TDS: 3.45–98.12 mg/L; TSS: 0.51–40.61 mg/L), temperature (18.70–25.90°C), and turbidity (0.41–9.77 NTU) and decreases the pH of rainwater quality.

5. Aluminum and zinc roofs are recommended for rainwater utilization but must be filtered and pre-treated before use, stone-coated and asbestos roofs are not the best surfaces for rainwater harvesting, and as such, rainwater from asbestos roofs should not be utilized due to their negative health implications on humans.

Rainwater is pure but due to the negative impact from contaminant release into the atmosphere, it interacts with rainwater thereby influencing the quality of rainwater portability. Rainwater collected from diverse sampling surfaces shows the satisfactory concentration in terms of the physicochemical properties of rainwater as dilution takes place from early rain to late rain, hence, further purification and treatment must be done to safeguard the health and well-being of humans. The negative impacts from anthropogenic releases into the atmosphere, which interacts with rainwater cause adverse health-related issues for man and the environment.

Acknowledgements

We would like to express our sincere gratitude to all those who have contributed to the completion of this research work, as their support, guidance, and encouragement have been invaluable in making this research possible. Also, we acknowledge anonymous reviewers and editors of this book series who were involved in the entire process.

Conflict of interest

All the authors declare no conflict of interest regarding this manuscript.

Notes/thanks/other declarations

We are thankful for the invitation to contribute to Rainfall - Observations and Modeling and other anonymous reviewers for their assistance. This research work was self-funded, as the authors declare no conflict of interest.

Author details

Daniel O. Omokpariola[1,2*], John K. Nduka[1] and Patrick L. Omokpariola[2]

1 Faculty of Physical Sciences, Department of Pure and Industrial Chemistry, Nnamdi Azikiwe University, Awka, Nigeria

2 Directorate of Chemical Evaluation and Regulation, National Agency for Food and Drug Administration and Control, Lagos, Nigeria

*Address all correspondence to: omeodisemi@gmail.com

References

[1] Omokpariola DO, Nduka JK, Omokpariola PL, Omokpariola ECO. Ionic composition of rainwater from different sampling surfaces across selected locations in Rivers state, Nigeria. World Scientific News. 2020; **150**:132-147

[2] UNESCO. Rainwater Harvesting: A Pathway to Enhance Resilience to Climate Change in Africa. Nairobi, Kenya: United Nations Educational, Scientific and Cultural Organization; 2019

[3] Ali S, Sang YF. Implementing rainwater harvesting systems as a novel approach for saving water and energy in flat urban areas. Sustainable Cities and Society. 2023;**89**:104304

[4] De Sá Silva ACR, Bimbato AM, Balestieri JAP, Vilanova MRN. Exploring environmental, economic, and social aspects of rainwater harvesting systems: A review. Sustainable Cities and Society. 2022;**76**:103475

[5] Teston A, Piccinini Scolaro T, Kuntz Maykot J, Ghisi E. Comprehensive environmental assessment of rainwater harvesting systems: A literature review. Water. 2022;**14**(17):2716

[6] Custódio DA, Ghisi E. Impact of residential rainwater harvesting on stormwater runoff. Journal of Environmental Management. 2023;**326**: 116814

[7] Silva MPD, González J, da Costa BB, Garrido C, Soares CA, Haddad AN. Environmental impacts of rainwater harvesting Systems in Urban Areas Applying Life Cycle Assessment—LCA. Engineering. 2023;**4**(2):1127-1143

[8] Nduka JKC, Orisakwe OE, Ezenweke LO, Ezenwa TE, Chendo MN, Ezeabasili NG. Acid rain phenomenon in Niger Delta region of Nigeria: Economic, biodiversity, and public health concern. The Scientific World Journal. 2008;**8**: 811-818

[9] Nduka JK, Orisakwe OE. Precipitation chemistry and occurrence of acid rain over the oil-producing Niger Delta region of Nigeria. The Scientific World Journal. 2010;**10**:528-534

[10] Nduka JKC, Orisakwe OE. Water quality issues in the Niger Delta of Nigeria: Polyaromatic and straight chain hydrocarbons in some selected surface waters. Water Quality Exposure and Health. 2010;**2**:65-74

[11] Ghose T, Chakraborty S, Ray R. Physico-chemical and bacteriological analysis of rainwater harvesting system in a metropolitan city. Water Resources and Industry. 2017;**17**:9-18

[12] Saalidong BM, Aram SA, Otu S, Lartey PO. Examining the dynamics of the relationship between water pH and other water quality parameters in ground and surface water systems. PLoS One. 2022;**17**(1):e0262117

[13] Liang L, Deng Y, Li J, Zhou Z, Tuo Y. Modelling of pH changes in alkaline lakes with water transfer from a neutral river. Chemosphere. 2023;**310**:136882

[14] Price GA, Stauber JL, Jolley DF, Koppel DJ, Van Genderen EJ, Ryan AC, et al. Natural organic matter source, concentration, and pH influences the toxicity of zinc to a freshwater microalga. Environmental Pollution. 2023;**318**:120797

[15] Okudo CC, Ekere NR, Okoye CO. Quality assessment of non-roof

harvested rainwater in industrial layouts of Enugu, Southeast Nigeria. Applied Water Science. 2023; **13**(5):1-9

[16] Santana TC, Guiselini C, Cavalcanti SDL, da Silva MV, Vigoderis RB, Júnior JAS, et al. Quality of rainwater drained by a green roof in the metropolitan region of Recife, Brazil. Journal of Water Process Engineering. 2022;**49**:102953

[17] Sampaio CA, Terezo RF, Ide GM, Spanholi CA, Matos FM, Burgardt T. Rainwater harvesting roofs: Insights of water quality and potential usage in rural areas. Revista de Ciências Agroveterinárias. Universidade do Estado de Santa Catarina. 2022;**21**(3): 354-359

[18] Omokpariola DO, Omokpariola PL. Health and exposure risk assessment of heavy metals in rainwater samples from selected locations in Rivers State, Nigeria. In: Green and Sustainable Processing. Physical Science Reviews. Vol. 2. Berlin, Germany. 2021. pp. 1-8

[19] Ashbolt NJ, Grabow WOK, Snozzi M. Indicators for Waterborne Pathogens. Geneva, Switzerland: World Health Organization; 2018

[20] Nduka JK, Omokpariola DO, Kelle HI, Iduseri EO, Mgbemena MN. Chemometric and risk assessment of nitrogen composition of atmospheric rainwater from diverse surfaces in Rivers state, Nigeria. Environmental Monitoring and Assessment. 2022; **194**(11):807

[21] Omokpariola DO, Nduka JK, Kelle HI, Mgbemena NM, Iduseri EO. Chemometrics, health risk assessment and probable sources of soluble total petroleum hydrocarbons in atmospheric rainwater, Rivers state, Nigeria. Scientific Reports. 2022;**12**(1):11829

[22] Omokpariola DO. Influence on storage condition and time on properties of carbonated beverages from utilization of polyethylene terephthalate (PET) bottles: Chemometric and health risk assessment. Environmental Analytical Health and Toxicology. 2022;**37**(3): e2022019

[23] Sreemahadavan-Pillai. A Comprehensive Laboratory Manual for Environmental Science and Engineering. New Delhi, India: New Age International (P) Ltd., Publishers; 2009 ISBN (13): 978-81-224-2951-0

[24] APHA. Standard methods for the examination of water and wastewater. 21st ed. Washington, DC, United States: American Public Health Association, American Water Works Association, Water Environment Federation; 1999

[25] USEPA. Drinking Water Microbiology Committee on Challenges of Modern Society. Wato/ccm Drinking Water Pilot Project Series. Edition by D. A. Uiver and R. A. Newman EPA, Washington D. C. 2007. Available from: www.unesco.org/water/newsletter/161.shtml [Accessed: March 2019]

[26] Omokpariola PL, Omokpariola DO. Institutional framework for the sound Management of Chemicals and Chemical Industries in Nigeria. ACS Chemical Health and Safety. 2021;**28**(6):457-467. DOI: 10.1021/acs.chas.1c00045

[27] Zaiontz C. XRealStats Real Statistics Using Excel. 2020. Available from: www.real-statistics.com

[28] Hammer Ø, Harper DA. Past: Paleontological statistics software package for education and data analysis. Palaeontologia Electronica. 2001;**4**(1):1

[29] Hammer Ø, Harper DA. Paleontological Data Analysis. New Delhi, India: John Wiley & Sons; 2008

[30] Panyakapo M, Onchang R. A four-year investigation on wet deposition in western Thailand. Journal of Environmental Sciences. 2008;**20**:441-448

[31] Kovacs J, Tanos P, Korponai J, Szekely IK, Gondar K, Gondar-Soregi K, et al. Analysis of Water Quality Data for Scientists. London, UK, London, UK: Intech Publishers; 2012. pp. 65-94 ISBN: 978-953-51-0486-5

[32] Das CR, Das S, Panda S. Groundwater quality monitoring by correlation, regression and hierarchical clustering analyses using WQI and PAST tools. Groundwater for Sustainable Development. 2022;**16**:100708

[33] Kasozi KI, Otim EO, Zirintunda G, Tamale A, Otim O. Multivariate analysis of heavy metals content of beef from Soroti, Uganda. Toxicology Reports. 2023;**10**:400-408

[34] Ojaniyi AF, Okoye PAC, Omokpariola DO. Heavy metals analysis and health risk assessment of three fish species, surface water and sediment samples in Ogbaru Axis of river Niger, Anambra state, Nigeria. Asian Journal of Applied Chemistry Research. 2021;**9**(1): 64-81

[35] Okpoebo UC, Jayeoye TJ, Adebayo AJ, Oguntimehin II. Environmental implications and significance of rainwater harvested from Lagos, Southwest Nigeria. Journal of Environmental Analytical Chemistry. 2014;**1**:118

[36] Gav BL, Vesuwe RN, Ijeomah AO, Okoko EE, Tor PN, Tsaviv JN. Determination of heavy metal concentrations of rainwater harvested from different roofing sheets in outskirt of Makurdi, Benue state, Nigeria. Asian Journal of Chemical Sciences. 2018;**4**(1): 1-6

[37] Al-Amoush H, Al-Ayyash S, Shdeifat A. Harvested rainwater quality of different roofing material types in water harvesting system at Al al-Bayt university/Jordan. Jordan Journal of Civil Engineering. 2018;**12**(2):228-244

[38] WHO. Guidelines for Drinking Water Quality. Recommendation. 4th ed. New Delhi: AUBS publishers; 2011a. p. 564. ISBN 978 92 4 154815 1

[39] Igbinosa IH, Aighewi IT. Assessment of the physicochemical and heavy metal qualities of rooftop harvested rainwater in a rural community. Global Challenges. Sep 2017;**1**(6):1700011. DOI: 10.1002/gch2.201700011

[40] Chaplot V, Selala MS, Thenga H, Jewitt GPW. Comparison of the chemical quality of rainwater harvested from roof and surface run-off systems. Water SA. 2018;**44**(2):223-231. DOI: 10.4314/wsa.v44i2.08

[41] WHO. Adverse Health Effects of Heavy metals in Children. Geneva, Switzerland: Children's Health and the Environment; 2011b

[42] Eruola AO, Ufoegbune GC, Eruola AO, Awomeso JA, Adeofun CO, Idowu OA, et al. Qualitative and quantitative assessment of rainwater harvesting from rooftop catchment: Case study of Oke-Lantoro community in Abeokuta, Southwest Nigeria. Journal of Agricultural Science and Environment. 2010;**10**(2):45-58

[43] Cape JN, Tang YS, Gonzalez-Benıtez JM, Mitosinkova M, Makkonen U, Jocher

M, et al. Organic nitrogen in precipitation across Europe. Biogeosciences. 2012;**9**:4401-4409. DOI: 10.5194/bg-9-4401-2012

[44] DeBusk K, Hunt WF. Rainwater Harvesting: A comprehensive review of literature. North Carolina, US: North Carolina Water Resources Research Institute; Project No. 11-12-W; 2012

[45] Okechukwu VU, Omokpariola DO, Onwukeme VI, Nweke EN, Omokpariola PL. Pollution investigation and risk assessment of polycyclic aromatic hydrocarbons in soil and water from selected dumpsite locations in Rivers and Bayelsa state, Nigeria. Environmental Analysis Health and Toxicology. 2021;**36** (4):e2021023. DOI: 10.5620/eaht.2021023

[46] Prasanna MV, Chidambaram S, Srinivasamoorthy K. Statistical analysis of the hydro-geochemical evolution of groundwater in hard and sedimentary aquifers system of Gadilam river basin, South India. Journal of King Saud University, Science. 2010;**22**: 133-145

[47] Khare P, Goel A, Patel D, Behari J. Chemical characterization of rainwater at a developing habitat of northern India. Atmospheric Environment. 2004;**69**: 135-145

[48] Uchiyama R, Okochi H, Katsui N, Ogata H. The impact of air pollutants on rainwater chemistry during "urban-induced heavy rainfall" in downtown Tokyo, Japan. Journal of Geophysical Research: Atmospheres. 2017;**122**:6502-6519. DOI: 10.1002/2017JD026803

[49] Ohara T. Study on characteristics of photochemical oxidants and particulate matter. Research Report from the National Institute for Environmental Studies. 2010;**203**:1-219

[50] Kumar P, Yadav S, Kumar A. Sources and processes governing rainwater chemistry in New Delhi, India. Natural Hazards. 2014;**74**:2147-2162

[51] Lü P, Han G, Wu Q. Chemical characteristics of rainwater in karst rural areas, Guizhou Province, Southwest China. Acta Geochimica. 2017;**36**:572-576

[52] Salve PR, Maurya A, Wate SR, Devotta S. Chemical composition of major ions in rainwater. Bulletin of Environmental Contamination and Toxicology. 2008;**80**(3):242-246

[53] Thepanodh S, Ayer GP, Hooper MA. Analysis of precipitation chemistry in northern Thailand. Clean Air and Environmental Quality. 2005;**39**:43-47

[54] UWINEZA A, IRIE M. Flood analysis for estimating the impact of rainwater harvesting system installation using hydrological models. Case study: Nyabugogo valley, Kigali. Journal of Arid Land Studies. 2022;**32**(S):145-149

[55] Heijerick DG, Janssen CR, Karlen C, Odnevall Wallinder I, Leygraf C. Bioavailability of zinc in runoff water from roofing materials. Chemosphere. 2002;**47**(10):1073-1080

[56] VM-Zinc. Composition of Zinc Roof. Available from: https://www.vmzinc.co.uk/zinc-basics/composition-of-zinc.html [Accessed: April 2021].

[57] United-America. Chemical Composition and Properties of Aluminium Roof. Available from: https://unitedaluminium.com/chemical-composition-and-properties-of-aluminium-alloys/ [Accessed: April 2021].

[58] Benton DM. Identification of Asbestos using laser-induced breakdown spectroscopy: A viable alternative to the

conventional approach? ISRN Spectroscopy. 2013;**2013**:1-6

[59] Dofasco. Steel Material Safety Data Sheets (MSDS). Available from: http://wcm.pavliks.com/WCMAdmin/Images/wwwbmp-groupcom/Customer_Service_Images/MSDSsheets.pdf [Accessed: August 2019].

[60] Omokpariola DO, Omokpariola ECO, Okechukwu VU. Simulation studies on corrosion of stone-coated roofing sheets sold in Nigeria. Bulletin in Chemical Society of Ethiopia. 2021;**35** (2):461-470. DOI: 10.4314/bcse.v35i2.18

[61] Simmons G, Hope V, Lewis G, Whitmore J, Gao W. Contamination of potable roof-collected rainwater in Auckland, New Zealand. Water Research. 2001;**35**(6):1518-1524

[62] Adeniyi IF, Olabanji IO. The physio-chemical and bacteriological quality of rainwater collected over different roofing materials in Ile-Ife, South-Western Nigeria. Chemical Ecology. 2005;**21**(3):149-166

[63] Mendez CB, Klenzendorf JB, Afshar BR, Simmons MT, Barrett ME, Kinney KA, et al. The effect of roofing material on the quality of harvested rainwater. Water Resources. 2011;**45**:2049-2059

[64] WHO. Guidelines for drinking water quality. Geneva, Switzerland: WHO Publications; 2008. Available from: http://www.whglibdoc.who.int/publications/2008/9241546964.eng.pdf [Accessed: April 2019]

Section 3

Impacts of Rainfall

Chapter 5

Rainfall's Symphony: Understanding Its Influence on Communication Systems in Nigeria

Obiseye Obiyemi and Katleho Moloi

Abstract

This study investigates the significant issue of rain-induced signal attenuation in satellite communication, specifically focusing on television receive-only (TVRO) stations operating in the Ku band across diverse Nigerian locations. Utilizing a comprehensive dataset of point rain rate distribution and the ITU rain attenuation model, the study comprehensively assesses how rainfall impacts signal quality. The findings highlight that southern regions consistently display high decibel (dB) values, indicating increased susceptibility to signal disruptions during heavy rainfall, while a comparative analysis between two key satellites, EUTELSAT 36B and INTELSAT 20, consistently favors the former in terms of signal resilience during adverse weather conditions. In contrast, northern regions generally exhibit lower dB values, suggesting a higher degree of signal resilience during rainfall events. These insights underscore the importance of considering location-specific and satellite-specific factors when designing satellite communication systems, emphasizing the need for robust infrastructure and strategic satellite selection to mitigate the effects of rain-induced attenuation. This study provides valuable guidance to engineers and service providers, enabling them to make informed decisions to minimize signal disruptions and enhance overall network reliability, particularly in regions susceptible to rain-induced attenuation.

Keywords: rain-induced attenuation, satellite communication, signal degradation, communication reliability, rainfall rates

1. Introduction

Rainfall can have a severe impact on the performance of microwave systems operating at frequencies above 10 GHz [1]. This phenomenon, known as rain attenuation or rain fade, results in the fading of propagating signals and can significantly degrade the quality and reliability of telecommunication systems during rainy periods. According to the studies conducted, rain attenuation has been identified as a major obstacle in the design of microwave communication links operating at higher frequencies [2, 3]. The attenuation caused by rainfall depends on various factors, such as the intensity of the rain, the size and shape of raindrops, and the distance between the transmitting and receiving antennas [4, 5]. The effect of heavy or intense rainfall is more pronounced in tropical and subtropical regions where the frequency and intensity of rain occurrences are higher [5–7].

In such regions, the presence of large raindrops and a higher rainfall intensity further exacerbate the problem and increase the severity of signal degradation. These large raindrops can cause significant scattering and absorption of microwave signals, leading to a decrease in signal strength and an increase in error rates. It has been observed that the size distribution of raindrops can vary depending on the geographical location, which in turn strongly influences the specific attenuation caused by rainfall and, consequently, the overall rain attenuation [8, 9]. Rainfall intensity, on the other hand, can also vary significantly depending on the location and meteorological conditions, further impacting the extent of signal degradation caused by rainfall [7, 10]. Furthermore, rainfall can also affect the polarization of the electromagnetic wave as it propagates through raindrops [5]. This change in polarization can further contribute to signal degradation and a reduction in the quality of the received signal. Moreover, the magnitude of rain attenuation is also influenced by other parameters, such as rain temperature, drop velocity, drop orientation, and transmitting frequency [7].

This is particularly relevant in areas with high rainfall rates, such as Nigeria, where excessive rainfall is a common occurrence throughout the year. Moreover, rainfall rates can vary significantly in small locations over short distances, making rain attenuation even more significant in causing propagation loss and affecting millimetric and microwave signals. Rain attenuation is primarily caused by two factors: scattering and absorption of the electromagnetic signal by water droplets in the atmosphere [11]. These factors lead to a reduction in signal strength and quality, resulting in outages, unavailability, and poor service performance. Several sources confirm that rainfall attenuation becomes significant and disruptive in satellite communications at frequencies above 10 GHz. The severity of rain attenuation is further exacerbated in tropical regions due to the heavy rainfall intensity and the presence of large raindrops, as shown in **Figure 1**. It is important to note that the size distribution of raindrops can vary depending on geographical location, which in turn strongly influences the rain-specific attenuation and overall rain attenuation. Rainfall attenuation has a direct impact on the performance of microwave

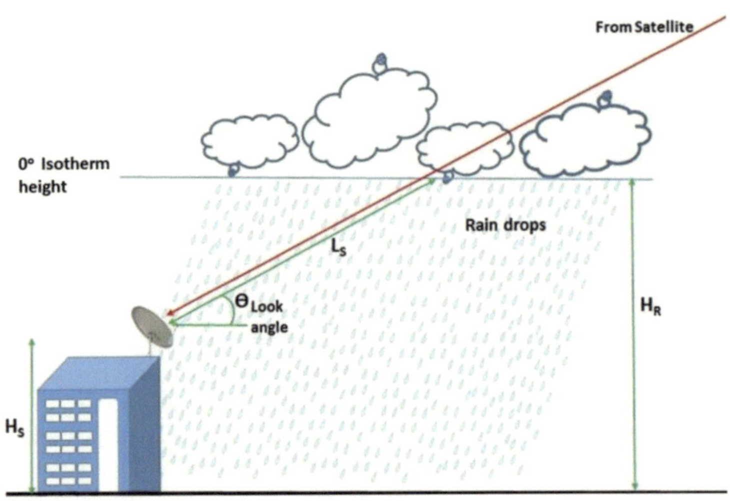

Figure 1.
Geometry of a typical rain-degraded satellite-earth link [12].

communication systems, limiting the scope of coverage and degrading system performance. Understanding the impact of rainfall on microwave signals is crucial for designing and maintaining reliable communication systems.

Figure 1 demonstrates that raindrops, in the context of typical Ku-band reception via the earth-space link, can result in the absorption, scattering, diffraction, and depolarization of electromagnetic signals. Satellite broadcast in the Ku-band (12–18 GHz) offers smaller antennas and less interference at higher frequencies than traditional terrestrial broadcast in the Very High Frequency (VHF) (30–300 MHz) and Ultra High Frequency (UHF) (300–3 GHz) bands. However, this frequency advantage is counterbalanced by a reduction in wavelength.

The average raindrop size (ARS) is defined by a diameter of 1.67 mm [4], but signals in the 12–18 GHz range exhibit wavelengths ranging from 25 to 16.7 mm. This implies that the wavelength of the signal is consistently greater than the typical size of rain droplets, which typically range in diameter from 0.1 to 5 mm. Nevertheless, despite the discrepancy in size, the signal encounters increased attenuation when rain is present. For frequencies up to 3 GHz, the droplet size in the Rayleigh scattering by raindrops is considerably lower than the wavelength. This function relates to the scattering characteristics of raindrops, which are affected by factors such as the size, shape, complex permittivity, and frequency of the signal [4].

Rainfall statistics and studies on rain-induced attenuation in Nigeria play a crucial role in the design and planning of reliable communication systems and infrastructure development in the country. Accurate knowledge of the rainfall rate and its impact on communication systems is essential for ensuring reliable transmission and minimizing signal loss due to rain-induced attenuation. Research efforts have been made in the past few decades to characterize the statistical and dynamic aspects of rainfall rate and rain attenuation in different regions of Nigeria. These studies have shed light on the variability and intensity of rainfall in different parts of the country and have provided valuable insights into the attenuation of radio waves during rainfall events.

For instance, studies conducted by [13] Ajayi in Ile-Ife and [14, 15] Ilorin have provided some insights into the rainfall characteristics and rain-induced attenuation in these locations. However, the availability of comprehensive rainfall data analysis remains limited in certain regions of Nigeria. Specifically, there is a lack of robust rainfall data analysis and studies on rain-induced attenuation covering several parts of Nigeria. This knowledge gap poses challenges for the planning and design of communication systems in this region, as rain-induced attenuation can significantly affect the reliability and performance of these systems. Obtaining accurate and reliable information on rain rate statistics and its impact on radio wave attenuation is vital for the effective design and planning of communication systems in this region. This study aims to evaluate the influence of rainfall-induced signal attenuation on satellite communication in various Nigerian regions, especially for digital satellite television as a typical application operating at frequencies above 10 GHz. By utilizing comprehensive rain rate data and the ITU rain attenuation model, the research seeks to assess the susceptibility of different locations and satellite systems to signal disruptions, facilitating the development of strategies to improve communication reliability during adverse weather conditions.

2. Related studies

Several studies have been conducted to investigate rain attenuation in Nigeria, which is a crucial factor affecting the quality of wireless communication systems.

These studies aim to understand the characteristics of rain attenuation, such as its frequency dependence, temporal and spatial variations, and its impact on different frequency bands used for wireless communication.

The research conducted in Nigeria has identified the important factors affecting rain attenuation, including frequency, rain rate, drop size distribution, drop shape, and atmospheric conditions, leading to a better understanding and characterization of rain attenuation in the region. For example, the study in Semire et al. [16] explores rainfall trends and variability in southwestern Nigeria, revealing a steady increase and slightly unstable volume variation, with higher attenuation in years due to increasing trend and frequency variability. The study by Omotosho and Oluwafemi [17] examines rainfall's impact on radio waves for fixed satellite service in Nigeria, finding practicable availability in the Ku, Ka, and V bands over different regions of the country. The study in [18] examines the impact of rainfall on digital satellite television reception in Akure, Nigeria, revealing signal squelching and complete loss at rainfall intensities above 64 mm/h.

The paper Ojo et al. [19] presents rain-rate contour maps for Nigerian link and network power sizing, developed using Moupfouma-Martin and Rice-Holmberg statistical methods for tropical regions. Obiyemi et al. [20] presents an experimental assessment of rain attenuation in Nigeria, analyzing cumulative rain rate distributions and fade margin estimation for various stations, revealing spatial variability. In a similar study on rain heights in satellite communication systems in Nigeria, the study by [21–23] reveals an increasing trend in Nigeria's rain heights, with ITU-R recommended values underestimating them, and suggests a preference for GPM data for estimating rain heights. This study by [24, 25] analyzes tropical rain data from the TRMM satellite to predict attenuation due to precipitation phenomena, revealing seasonal, rainfall type dependence, and overestimation of rain heights in Nigeria.

The study by Ojo et al. [26] examines the influence of weather factors, notably rain-induced attenuation and various hydrometeors, on 5G satellite communication channels. The implementation of a dynamic adaptive intelligent scheme (DAIS) using fuzzy logic demonstrates noteworthy enhancements in service quality, achieving a reduction in power and signal-to-noise ratio (SNR) demands without compromising data integrity. These findings hold notable relevance for satellite and wireless network performance, especially within the context of Nigeria's adoption of digital video broadcasting (DVB).

The study by Ojo et al. [27] presents contour maps as useful tools for rain rate and rain attenuation prediction, utilizing 30 years of rainfall data in Nigeria. These maps have been useful in designing microwave links, offering valuable insights for engineers. Based on data from ground-based precipitation measurements in Nigeria, Ojo [28] highlights the relationship between co-polar attenuation, cross-polar discrimination (XPD), and rainfall rates across different elevation angles and frequencies. Using a 10-year weather dataset for locations in Northern Nigeria, Orji et al. [29] examines the impact of rain on signal strength over the Ku and V bands, with results showing significant effects in V-band frequencies. On experimental efforts on the quantification of the attenuation induced on the typical Ku satellite-earth link, efforts documented in the studies in Refs. [18, 30–32] provide pronounced evidence on the degrading impacts of rain on digital satellite television applications over the region. Studies in Ref.s [33, 34] explore rain rate and attenuation across Nigeria's climatic zones, emphasizing the worst months' impact on earth-space radio links and presenting useful statistics such as the Average Worst Month and Average Yearly Worst Month for communication link designs over specified locations across Nigeria.

The current state of research on rainfall and its associated signal attenuation in Nigeria is marked by a notable gap in knowledge [33], which has significantly been influenced by the dearth of data. Addressing this gap is, however, imperative in order to enhance the reliability of communication networks across the country's diverse climatic regions.

3. Methodology

3.1 Nigerian climatology

Nigeria boasts an extensive range of geographical features and climates due to its strategic location between 4 and 14 degrees north latitude and 2 and 15 degrees east longitude in Western Africa. This positioning exposes the country to a rich tapestry of environmental diversity. The climate of Nigeria is intricately shaped by the dynamic seasonal movements of the intertropical discontinuity (ITD). During the wet or rainy season, which typically spans from April to October, Nigeria experiences the dominance of moist south-westerly winds originating from the South Atlantic Ocean. These winds carry a vital payload of moisture, crucial for sustaining the nation's rainfall patterns and frequent thunderstorms [25].

Conversely, the dry season, often referred to as Harmattan, extends from November to March. This period witnesses the north-easterly migration of dust particles from the vast Sahara Desert, sweeping across the entire country. These annual climatic shifts are driven by a combination of factors, including fluctuating temperatures, variations in precipitation levels, and other meteorological phenomena. Thanks to its equatorial location, Nigeria is blessed with a consistently high percentage of daylight year-round, contributing to its unique and diverse climate. This diverse climatic mosaic not only influences the environment but also plays a significant role in shaping the country's culture, agriculture, and way of life.

3.2 Rain attenuation estimates on digital satellite TV links

A comprehensive dataset detailing the distribution of point rain rate (R0.01 in mm/h) across specific locations across Nigeria, as documented in studies [17, 35] in the references, was employed for this study. The dataset includes point rainfall rates from 37 stations in Nigeria between 1998 and 2006, as derived from the Tropical Rainfall Measuring Mission (TRMM) satellite sensors, specifically the Microwave Imager (TMI, 3A12 V6) and other satellite sources (3B43 V6) as outlined in Ref. [17]. The calculations involved deriving total rain accumulation (M) and point rainfall rates, with estimates aligning well with comparable data from the ITU-R database and in-situ ground-based data for the same locations. Rainfall rates were computed based on regional climatic parameters, including the thunderstorm ratio, using the method derived by the study [36] in the references, and the average total rain deduced per year is presented in **Figure 2** for the observation period.

The study focuses on the earth station endpoints of the communication links, specifically television receive-only (TVRO) stations, as depicted in **Figure 3** for 37 satellite-earth links in the Ku band. This figure illustrates the precise locations of these selected sites in relation to the existing rain rate distribution across Nigeria. It's important to note that our selection process for these sites was deliberate, with the goal of ensuring a balanced representation of Nigeria's diverse physical and climatic

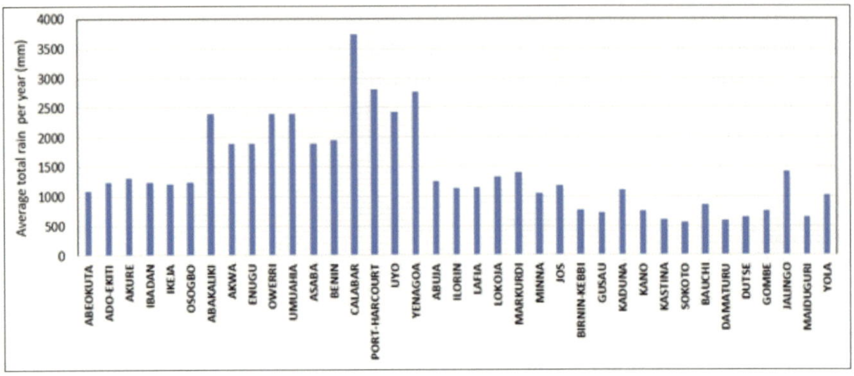

Figure 2.
Average total rain per year (mm) estimated from rain events in Nigeria as derived from TRMM 3B43V6 data: January 1998–December 2006 [17].

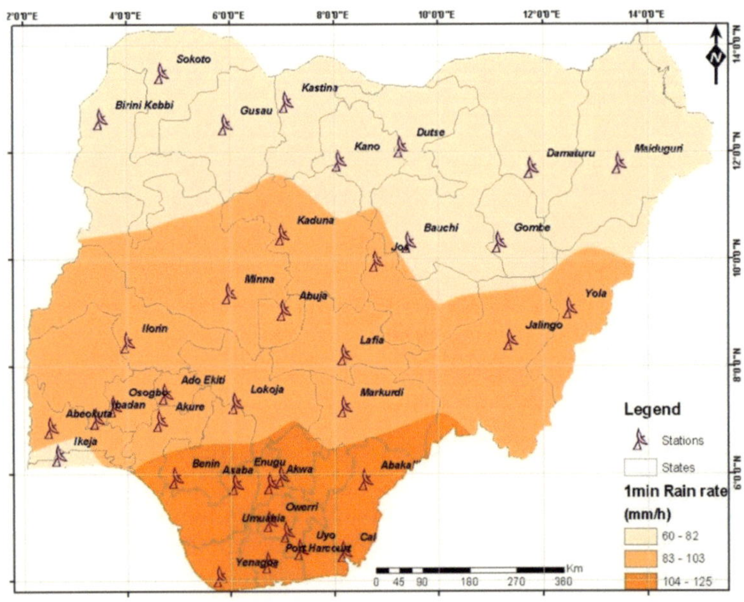

Figure 3.
TV reception stations (TVRO) spread across Nigeria with point rain rate distribution.

characteristics. Each chosen site was carefully considered to capture the unique geographical and environmental features of different regions within the country.

The two Ku transponders considered have a robust footprint over the region and prominently deliver satellite TV content across the region. They are EUTELSAT 36B with an orbital position of 36°E, an effective isotropic radiated power (EIRP) of 48 dBW, and a transmit frequency of 12.245 GHz, while the second one is INTELSAT 20 with an orbital position of 68.5°E, an EIRP of 48.5 dBW, and a transmit frequency of 12.722 GHz. One crucial factor in predicting rain attenuation is the elevation angle, and this parameter is influenced by two key factors: the orbital position of the satellites considered in the communication system and the geographic coordinates (latitude and longitude) of the earth station. Consequently, the look angle, which plays a vital role in signal propagation, exhibits significant variations across our designated communication links and from site to site, adding complexity to our analysis.

To construct a comprehensive rain attenuation map for Nigeria, the ITU rain attenuation model, detailed in the recommendation [37], was employed for quantifying the rain-induced attenuation over the links for each Ku transponder. This model relies on specific input parameters, including point rainfall rate (0.01% of an average year in mm/h) at the location, Earth station altitude above sea level (in km), elevation angle, latitude of the Earth station (in degrees), operating frequency (in GHz), and the standardized effective Earth radius of 8500 km. These parameters collectively enable a precise assessment of rain-induced signal attenuation across Nigeria's diverse geographical and climatic regions, enhancing the robustness of the study.

4. Results and discussions

The rain attenuation contour maps offer valuable insights into the susceptibility of various Nigerian locations to rain-induced signal degradation, with a specific focus on two satellites, EUTELSAT 36B and INTELSAT 20, represented in decibels (dB). Based on the estimates in **Figures 4** and **5** for rain-induced attenuation of 0.001% of

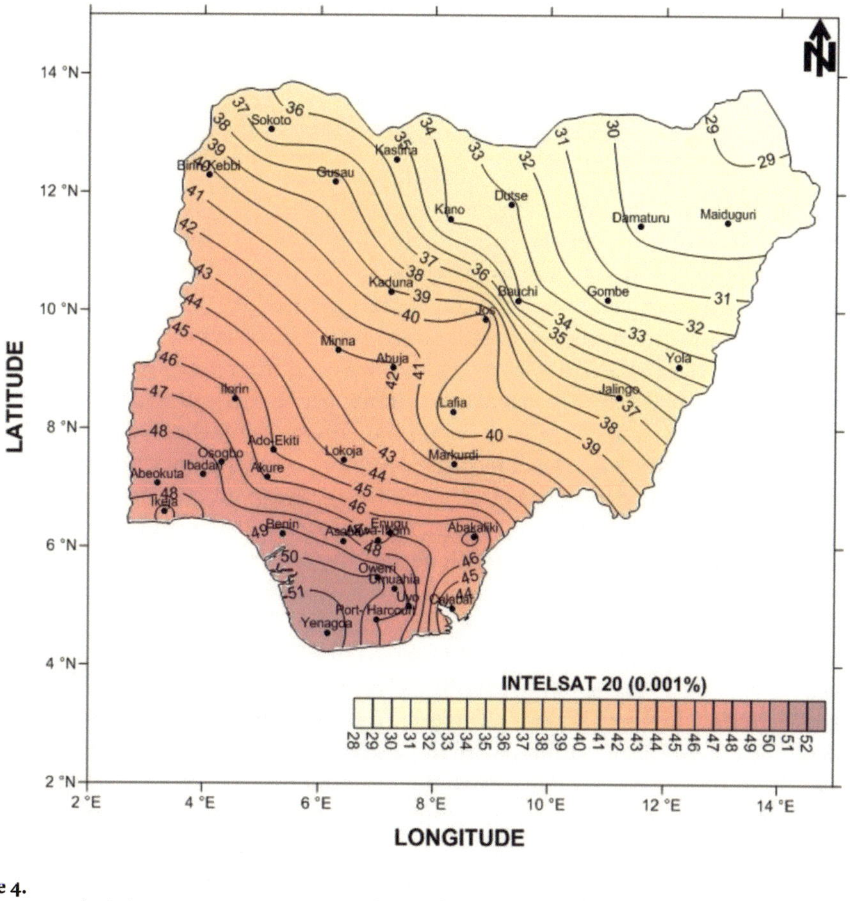

Figure 4.
A contour map depicting 0.001% rain attenuation (in dB) for typical Ku-band links on INTELSAT 20.

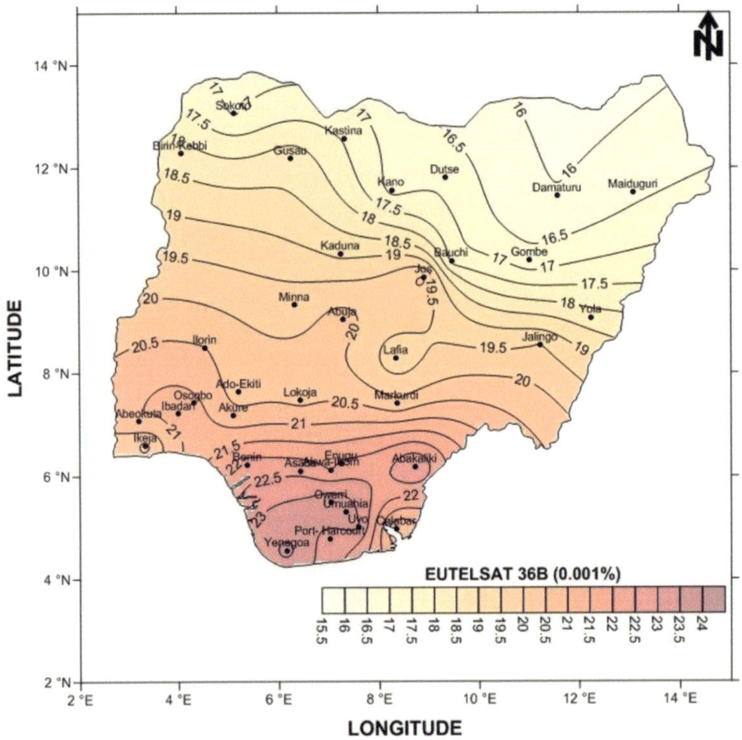

Figure 5.
A contour map depicting 0.001% rain attenuation (in dB) for typical Ku-band links on EUTELSAT 36B.

the time, southern regions, including Yenagoa, Umuahia, and Owerri, consistently exhibit high dB values, ranging from approximately 49 to 52 dB for INTELSAT 20, indicating their heightened vulnerability to signal disruptions during heavy rainfall events. These areas experience substantial signal attenuation, underlining the need for robust communication infrastructure to ensure reliable services.

Furthermore, the comparison between the two satellites reveals a consistent pattern, with INTELSAT 20 consistently reporting higher dB values compared to EUTELSAT 36B across the considered locations. This suggests that INTELSAT 20 is more susceptible to signal disruptions caused by rain, emphasizing the importance of satellite selection in mitigating the effects of adverse weather conditions. Conversely, northern regions like Kano, Sokoto, and Maiduguri generally exhibit lower dB values, indicating a higher degree of signal resilience during rainfall events.

The estimates in **Figures 6** and 7, expressed in decibels (dB), provide valuable insights into the impact of rainfall on signal quality for various Nigerian locations and two satellites, EUTELSAT 36B and INTELSAT 20, focusing on the 0.01% percentage of the time. Southern regions, including Uyo, Yenagoa, Owerri, and Umuahia, consistently report higher dB values, ranging from approximately 30 to 31.52 dB for INTELSAT 20. These figures underscore the susceptibility of these areas to significant signal attenuation during rainfall events, highlighting the critical need for robust communication infrastructure to ensure reliable services in these regions.

Furthermore, the satellite comparison reveals that INTELSAT 20 consistently records higher dB values than EUTELSAT 36B across the considered locations. For instance, in Yenagoa, INTELSAT 20 demonstrates a dB value of 31.52, while

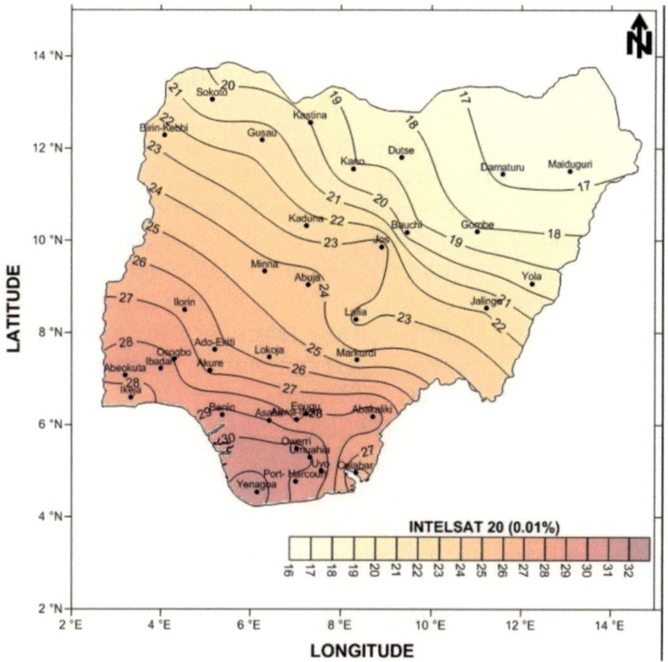

Figure 6.
A contour map depicting 0.01% rain attenuation (in dB) for typical Ku-band links on INTELSAT 20.

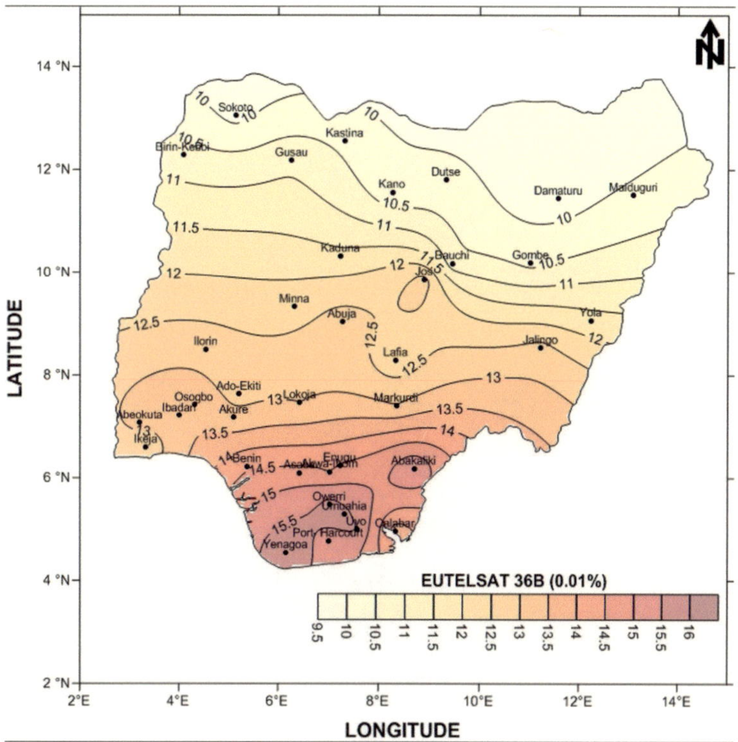

Figure 7.
A contour map depicting 0.01% rain attenuation (in dB) for typical Ku-band links on Eutelsat 36B.

EUTELSAT 36B records 15.84 dB. This consistent pattern indicates that INTELSAT 20 is more susceptible to signal disruptions caused by rain compared to EUTELSAT 36B. Conversely, northern regions such as Kano, Sokoto, and Maiduguri generally exhibit lower dB values, ranging from approximately 16.69 to 20.15 dB for INTELSAT 20, highlighting their greater signal resilience during rainfall events.

The rain-induced attenuation figures in **Figures 8** and **9**, expressed in decibels (dB) for the 0.1% time interval, provide further evidence of the significant influence of rainfall on signal quality in various Nigerian locations. Southern regions such as Uyo, Yenagoa, Owerri, and Umuahia consistently exhibit higher dB values, ranging from approximately 2.72 to 3.16 dB for INTELSAT 20. This reiterates the heightened vulnerability of these areas to substantial signal attenuation during rainy conditions, emphasizing the imperative need for robust communication infrastructure to ensure reliable services in the southern region of Nigeria.

Moreover, the persistent trend of INTELSAT 20 recording higher dB values compared to EUTELSAT 36B across the studied locations underscores the satellite's susceptibility to signal disruptions induced by rainfall. For instance, in Yenagoa, INTELSAT 20 registers a dB value of 3.16, while EUTELSAT 36B reports 1.38 dB. This consistent pattern reinforces the importance of considering satellite-specific characteristics when designing satellite communication systems to mitigate the effects of adverse weather conditions and enhance overall network reliability.

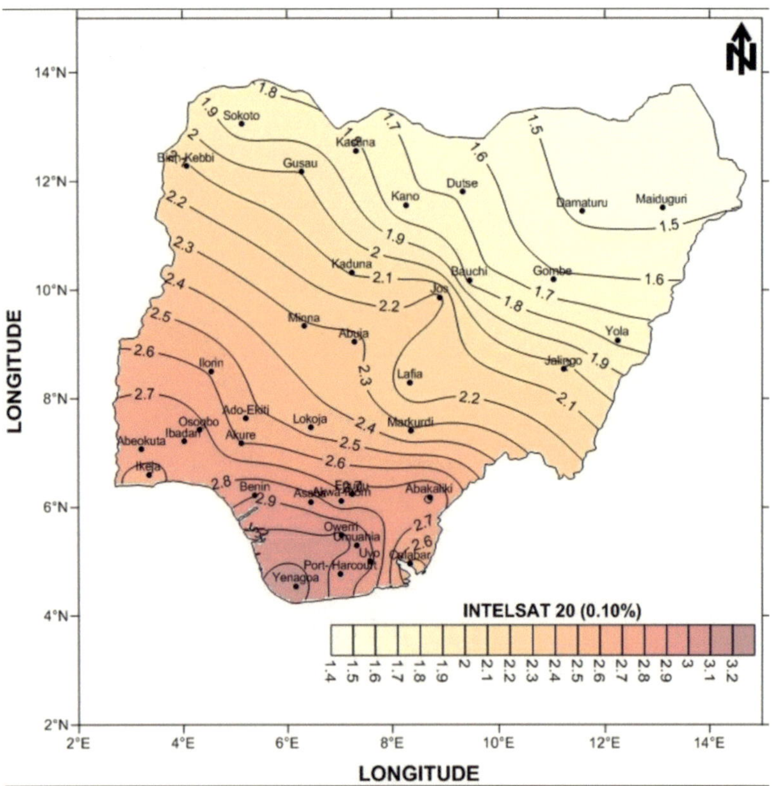

Figure 8.
A contour map depicting 0.1% rain attenuation (in dB) for typical Ku-band links on INTELSAT 20.

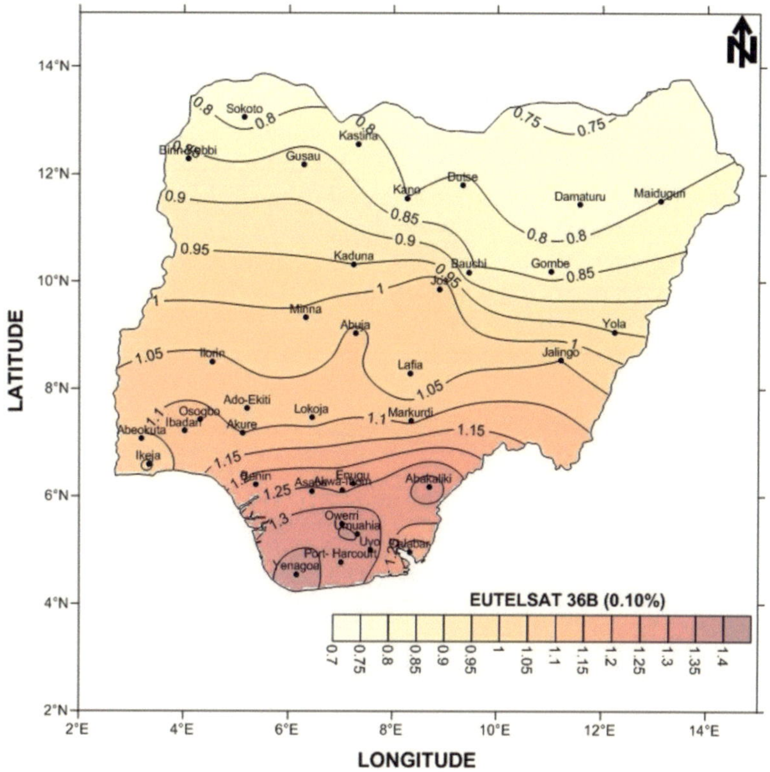

Figure 9.
A contour map depicting 0.1% rain attenuation (in dB) for typical Ku-band links on Eutelsat 36B.

In summary, these findings underscore the importance of considering both location-specific and satellite-specific rain-induced attenuation figures when optimizing satellite communication systems in Nigeria. High dB values signify areas prone to signal disruptions during rain, necessitating tailored solutions for reliable communication, while lower dB values indicate regions with higher signal resilience, aiding in the strategic deployment of communication infrastructure. These findings enable engineers and service providers to make informed decisions when optimizing satellite communication systems, particularly in regions with elevated rain-induced attenuation, to enhance network reliability and minimize signal disruptions. As observations and modeling studies consistently demonstrate a rise in heavy rains in various areas [38, 39], it is becoming evident that the future impact of climate change on communication systems will necessitate the development of more adaptive techniques to minimize signal disruptions at higher frequencies.

5. Conclusion

In conclusion, this research employed a comprehensive dataset of point rain rate distribution across various Nigerian locations to investigate rain-induced signal attenuation in satellite communication. The study focused on television receive-only (TVRO) stations in the Ku band, representing crucial endpoints of satellite communication links. The selection of these sites was strategic, ensuring a balanced representation of Nigeria's diverse geographical and climatic features.

The findings revealed the significant impact of rainfall on signal quality, with southern regions consistently demonstrating high dB values, indicating heightened vulnerability to signal disruptions during heavy rainfall events. The comparison between two prominent satellites, EUTELSAT 36B and INTELSAT 20, consistently favored the former in terms of signal resilience during adverse weather conditions. Conversely, northern regions generally exhibited lower dB values, suggesting greater signal resilience during rainfall events.

These insights underscore the importance of considering both location-specific and satellite-specific factors when designing satellite communication systems. Robust infrastructure and strategic satellite selection are essential to mitigate the effects of rain-induced attenuation, ensuring reliable communication services across Nigeria's diverse regions.

Conflict of interest

The authors declare no conflict of interest.

Author details

Obiseye Obiyemi[1*] and Katleho Moloi[2]

1 Space Science Centre, Department of Electrical Power Engineering, Durban University of Technology, Durban, South Africa

2 Department of Electrical Power Engineering, Durban University of Technology, Durban, South Africa

*Address all correspondence to: obiseyeo@dut.ac.za

References

[1] Chebil J, Zyoud A-H, Habaebi MH, Rafiqul IM, Dao H. Analysis of rain fade slope for terrestrial links. Indonesian Journal of Electrical Engineering and Computer Science. 2020;**18**:896-902

[2] Hudiono H, Taufik M, Perdana RHY, Rakhmania AE. Telemetering of rainfall measurement results using 433 MHz wireless transmission. Journal Infotel. 2021;**13**(3):143-150

[3] Kestwal MC, Joshi S, Garia LS. Prediction of rain attenuation and impact of rain in wave propagation at microwave frequency for tropical region (Uttarakhand, India). International Journal of Microwave Science and Technology. 2014;**2014**:958498

[4] Alozie E et al. A review on rain signal attenuation modeling, analysis and validation techniques: Advances, challenges and future direction. Sustainability. 2022;**14**(18):11744

[5] Akinyemi GA, Falade JA, Kolawole LB. Estimation of rain attenuation at microwave bands in Nigeria. Журнал радиоэлектроники. 2018;**8**:3

[6] Chiang L-C, Wang Y-C, Liao C-J. Spatiotemporal variation of sediment export from multiple Taiwan watersheds. International Journal of Environmental Research and Public Health. 2019;**16**(9):1610

[7] Iheanyichukwu CA, Ike OA, Chukwuemeka CP. Investigating the impact of weather parameters on signal strength of satellite dish in Enugu metropolis. International Journal of Astrophysics and Space Science. 2020;**8**(2):11

[8] Oluwole O, Nicolae G, Olawale O, Oludele A. Mobile virtual laboratory in Nigeria. International Journal of Engineering and Computer Science. 2015;**4**:11417-11421

[9] Ojo JS, Akoma DB, Adetan O, Falodun SE. Analysis of rain-induced attenuation at Ku-V band using gamma drop size distribution model over a tropical region. Wireless Personal Communications. 2022;**124**:1-20

[10] Akobre S, Ibrahim M, Salifu AM. Rain rate and rain attenuation geographical map for satellite system planning in Ghana. International Journal of Computer Applications. 2020;**177**:34-45

[11] Adetan O, Oludare Fagbohun O. Analytical approach to critical diameters in raindrop size distribution in Durban, South Africa. Asian Journal of Advanced Research and Reports. 2018;**1**(1):1-8

[12] Obiyemi Obiseye O, Oluwumi A, Ibiyemi Tunji S. Recent update on one-minute rainfall rate measurements for microwave applications in Nigeria. International Journal of Wireless and Mobile Computing. 2015;**3**(3):33-39

[13] Ajayi GO, Ofoche EBC. Some tropical rainfall rate characteristics at Ile-Ife for microwave and millimeter wave applications. Journal of Applied Meteorology and Climatology. 1984;**23**(4):562-567

[14] Bryant GH, Adimula I, Riva C, Brussaard G. Rain attenuation statistics from rain cell diameters and heights. International Journal of Satellite Communications. 2001;**19**(3):263-283

[15] Adimula IA, Falaiye O, Willoughby A. Effects of rain on

microwave and satellite communications in equatorial and tropical regions. Nigerian Journal of Physics. 2005;**17**(1):66-71

[16] Semire FA, Adekunle AJ, Abolade RO. Rainfall variability and its effect on communication link in southwestern Nigeria. International Journal of Big Data Mining for Global Warming. 2020;**2**(1):2050004

[17] Omotosho TV, Oluwafemi CO. One-minute rain rate distribution in Nigeria derived from TRMM satellite data. Journal of Atmospheric and Solar-Terrestrial Physics. 2009;**71**(5):625-633. DOI: 10.1016/j.jastp.2009.02.003

[18] Obiyemi OO, Ibiyemi TS. Experimental investigation of rainfall effect on digital satellite television reception in Nigeria: Initial results. In: 2014 IEEE 6th International Conference on Adaptive Science & Technology (ICAST). New York: Institute of Electrical and Electronics Engineers; 2014. pp. 1-4

[19] Ojo JS, Ajewole MO, Emiliani LD. One-minute rain-rate contour maps for microwave-communication-system planning in a tropical country: Nigeria. IEEE Antennas and Propagation Magazine. 2009;**51**(5):82-89

[20] Obiyemi OO, Omotosho TV, Oguntuase VA, Tijani I. Experimental assessment of cellular mobile performance along the railway corridor. Covenant Journal of Information and Communication Technology. 2015;**3**(2):12-22

[21] Lawal YB, Ojo JS, Falodun SE. Rain height statistics from GPM data for satellite communication systems in Nigeria. IOP Conference Series: Earth and Environmental Science. 2021;**655**(1):12038

[22] Lawal YB, Ojo JS, Falodun SE. Variability and trends in rain height retrieved from GPM and implications on rain-induced attenuation over Nigeria. Heliyon. 2021;7(10):e08108

[23] Lawal YB, Falodun SE, Ojo JS. Temporal evolution of atmospheric parameter-profiling on rain height over two geoclimatic regions in Nigeria. Journal of Atmospheric and Solar-Terrestrial Physics. 2020;**211**:105482. DOI: 10.1016/j.jastp.2020.105482

[24] Ojo JS, Ojo OL, Joseph-Ojo CI. Rain height information from TRMM precipitation radar and micro rain radar for radio communication studies in Nigeria. In: 2014 IEEE 6th International Conference on Adaptive Science & Technology (ICAST). New York: Institute of Electrical and Electronics Engineers; 2014. pp. 1-5. DOI: 10.1109/ICASTECH.2014.7068104

[25] Ojo JS. Rain height statistics based on 0 C isotherm height using TRMM precipitation data for earth-space satellite links in Nigeria. International Scholarly Research Notices. 2014;**2014**:798289

[26] Ojo JS, Olurotimi EO, Obiyemi OO. Assessment of Total attenuation and adaptive scheme for quality of service enhancement in tropical weather for satellite networks and 5G applications in Nigeria. Journal of Microwaves, Optoelectronics and Electromagnetic Applications. 2021;**20**:228-247

[27] Ojo J, Ajewole M, Sarkar S. Rain rate and rain attenuation prediction for satellite communication in Ku and Ka bands over Nigeria. Progress In Electromagnetics Research B. 2008;**5**:207-223

[28] Ojo JS. Estimation of cross-polarization due to rain over some stations in Nigeria.

Annals of Telecommunications des télécommunications. 2012;**67**(5-6):241-245

[29] Orji PO, Chiemeka IU, Obiegbuna DC, Eze CM. Estimation of rain attenuation effect on radio wave propagation for broadband communication over Northern Nigeria. IOP Conference Series: Earth and Environmental Science. 2023;**1178**(1):12027

[30] Ajewole MO, Durodola OM, Ojo JS. Some aspects of rain effects on the performance of Ku-band satellite signals in Akure, Nigeria. In: 2014 XXXIth URSI General Assembly and Scientific Symposium (URSI GASS). Vol. 2014. Belgium: International Union of Radio Science; pp. 1-4

[31] Ometan OO, Omotosho TV, Adewusi MO, Akinwumi SA, Emetere ME, Boyo AO. Validation of rain attenuation models with experimental data at Covenant University, southwest, Nigeria. International Journal of Engineering Research and Technology. 2019;**12**:2111-2116

[32] Gbenga-Ilori AO. Effect of atmospheric impairments on ku-band free-to-air digital satellite television signals in Lagos state. Nigerian Journal of Technology. 2020;**39**(1):238-245

[33] Ibitokun J, Ojo JS, Ogunjo ST. Comparative study of worst month rain rate and attenuation statistics for satellite links in Nigeria. Radio Science. 2023;**58**(1):e2022RS007557

[34] Osita I, Nymphas EF. Characteristics of worst hour rainfall rate for radio wave propagation modelling in Nigeria. Meteorology and Atmospheric Physics. 2019;**131**:251-261

[35] Obiyemi OO, Afullo TJ, Ibiyemi TS. Evaluation of rain degraded digital satellite television reception in tropical regions. International Journal of Scientific and Engineering Research. 2013;**4**(12):790-799

[36] Ito C, Hosoya Y. The thunderstorm ratio as a regional climatic parameter: Its effects on different integration time rain rate conversion, rain attenuation, site-diversity and rain depolarization. In: URSI Proceeding, GA02 Pap. Vol. 181. Belgium: International Union of Radio Science; 2002. p. 2002

[37] ITU-RP. 618-13. Propagation Data and Prediction Methods Required for the Design of Earth-Space Telecommunications Systems. Geneva: Recomm. P. 618-13, P Ser. ITU-R, Int. Telecomm. Union; 2017

[38] Li Y et al. Strong intensification of hourly rainfall extremes by urbanization. Geophysical Research Letters. 2020;**47**(14):e2020GL088758

[39] Panthou G et al. Rainfall intensification in tropical semi-arid regions: The Sahelian case. Environmental Research Letters. 2018;**13**(6):64013